G000168821

Multidimensional Databases and Data Warehousing

Copyright © 2010 by Morgan & Claypool

All rights reserved. No part of this publication may be reproduced, stored in a retrieval system, or transmitted in any form or by any means—electronic, mechanical, photocopy, recording, or any other except for brief quotations in printed reviews, without the prior permission of the publisher.

Multidimensional Databases and Data Warehousing

Christian S. Jensen, Torben Bach Pedersen, and Christian Thomsen

www.morganclaypool.com

ISBN: 9781608455379   paperback
ISBN: 9781608455386   ebook

DOI 10.2200/S00299ED1V01Y201009DTM009

A Publication in the Morgan & Claypool Publishers series
SYNTHESIS LECTURES ON DATA MANAGEMENT

Lecture #9
Series Editor: M. Tamer Özsu, *University of Waterloo*
Series ISSN
Synthesis Lectures on Data Management
Print 2153-5418   Electronic 2153-5426

Synthesis Lectures on Data Management

Editor

M. Tamer Özsu, *University of Waterloo*

Synthesis Lectures on Data Management is edited by Tamer Özsu of the University of Waterloo. The series will publish 50- to 125 page publications on topics pertaining to data management. The scope will largely follow the purview of premier information and computer science conferences, such as ACM SIGMOD, VLDB, ICDE, PODS, ICDT, and ACM KDD. Potential topics include, but not are limited to: query languages, database system architectures, transaction management, data warehousing, XML and databases, data stream systems, wide scale data distribution, multimedia data management, data mining, and related subjects.

Privacy-Preserving Data Publishing: An Overview
Raymond Chi-Wing Wong and Ada Wai-Chee Fu
2010

Keyword Search in Databases
Jeffrey Xu Yu, Lu Qin, and Lijun Chang
2009

Multidimensional Databases and Data Warehousing

Christian S. Jensen
Aarhus University, Denmark

Torben Bach Pedersen
Aalborg University, Denmark

Christian Thomsen
Aalborg University, Denmark

SYNTHESIS LECTURES ON DATA MANAGEMENT #9

MORGAN & CLAYPOOL PUBLISHERS

ABSTRACT

The present book's subject is multidimensional data models and data modeling concepts as they are applied in real data warehouses. The book aims to present the most important concepts within this subject in a precise and understandable manner.

The book's coverage of fundamental concepts includes data cubes and their elements, such as dimensions, facts, and measures and their representation in a relational setting; it includes architecture-related concepts; and it includes the querying of multidimensional databases. The book also covers advanced multidimensional concepts that are considered to be particularly important. This coverage includes advanced dimension-related concepts such as slowly changing dimensions, degenerate and junk dimensions, outriggers, parent-child hierarchies, and unbalanced, non-covering, and non-strict hierarchies.

The book offers a principled overview of key implementation techniques that are particularly important to multidimensional databases, including materialized views, bitmap indices, join indices, and star join processing. The book ends with a chapter that presents the literature on which the book is based and offers further readings for those readers who wish to engage in more in-depth study of specific aspects of the book's subject.

KEYWORDS

multidimensional database, multidimensional model, data warehouse, data cube, dimension, dimension hierarchy, multidimensional fact and measure, star schema, relational OLAP, SQL, MDX, complex dimension, materialized view, bitmap index, join index

Contents

Preface

The objective of the book is to offer university students, researchers, and practitioners a comprehensive, integrated, and understandable coverage of multidimensional data models and data modeling concepts.

While many parts of the book can be read with a general computer science background, e.g., as obtained through a couple of years of computer science studies, the book should be entirely readable by students who have passed the kind of introductory course on database management systems that is found in most undergraduate computer science curricula.

The book's target audience includes the following:

- M.Sc. students who wish to better understand the subject of data warehousing and would like to obtain a good foundation for studies of data mining. To facilitate the use of the book in university curricula, each chapter includes exercises.

- Ph.D. students and researchers who wish to quickly obtain an overview of the key concepts in multidimensional data modeling, e.g., as background for their own research.

- Data management practitioners who seek a concise, readable, and product-independent coverage of the subject area. We hope that the book will serve as an introduction to the concepts and benefits of multidimensional databases, which are well-known in the database community by now, to a much broader community.

In writing the book, we aimed to optimize for the number of concepts per page while being reasonably precise, concrete, and understandable. In comparison to research papers that often cover modeling aspects as a means towards another end or go into depth with a narrow or as-of-yet exotic modeling subject, we offer a broader coverage than any individual paper and a less fragmented coverage than what may be obtained from studying a collection of papers. We also avoid the typical reliance on formalization found in many research papers, to maintain focus on concepts rather than notation.

The book also differs from other available books. Database textbooks either do not cover multidimensional data modeling or, at best, include a brief coverage of the subject—this book offers a much more extensive coverage. Books on data warehousing that target practitioners come in at least two flavors. Some books are specific to particular database management systems or tools. In contrast, the present book is designed to be product independent and to offer a foundation for understanding a broad range of available systems. Other practitioner-oriented books also adopt a product-independent approach. However, all the practitioner-oriented books tend to have many pages and lots of screen shots or other figures, resulting in fewer concepts per page.

The book is organized as follows: Chapter 2 introduces fundamental concepts. Chapter 3 presents more advanced concepts. Chapter 4 covers the implementation of multidimensional database technology. Chapter 5 gives pointers to papers and books about the multidimensional data models.

The introduction and Chapters 2 and 3 cover modeling aspects. Chapter 4 is separate in that it covers implementation aspects. The dependency diagram in Figure 1 presents "recommended dependencies" at the section level for those readers who wish to read selected parts of the books.

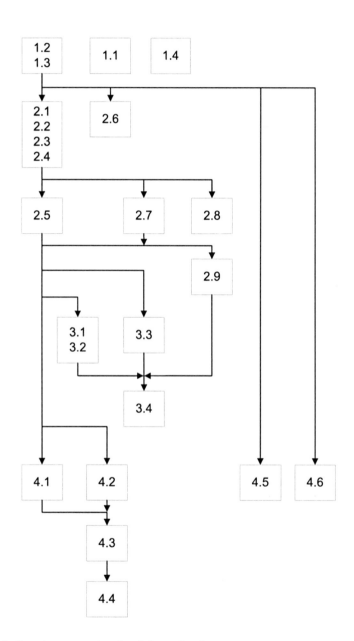

Figure 1: Reader's directions—section-level dependencies

Acknowledgments

Several individuals provided us with valuable assistance when we prepared this lecture. We thank Tamer Özsu for inviting us to contribute the lecture, and we thank Diane Cerra for her insightful management of the entire process. We thank both Tamer and Diane for helping us prioritize our time so that a lecture would eventually materialize.

We also thank Tamer for identifying problems such as typos and hard-to-follow prose in an earlier draft of the lecture, and for challenging us to find ways of using the noun "data" that comply with the editorial policy and our own preferences. Likewise, we thank LeeAnn Iovanni for proofreading parts of a near-final draft. We thank Artur Baniukevic for carrying out a variety of Latex editing tasks, and we thank the production team headed by C. L. Tondo for assistance.

Christian S. Jensen, Torben Bach Pedersen, and Christian Thomsen
August 2010

CHAPTER 1

Introduction

This first chapter sets the stage for the remainder of the book. It starts by giving a brief characterization of multidimensional data management and data modeling, then covers briefly the history of multidimensional data management and defines key terms to be used throughout the book. Then it explains why related technologies, specifically, spreadsheets and relations, fall short in meeting the modeling needs of multidimensional data management. Finally, it contrasts multidimensional data management with traditional relational data management.

1.1   MULTIDIMENSIONAL DATA MANAGEMENT

The relational data model, which was introduced by Codd in 1970 and earned him the Turing Award a decade later, constitutes a significant part of the foundation of today's multi-billion-dollar database industry. During the 1990s, a new type of data model, the *multidimensional data model*, emerged that has since made inroads on the relational model when the objective is to *analyze* data, rather than to perform on-line transactions. The multidimensional data model underpins the multi-billion-dollar business intelligence industry, and it plays a role in this industry that is similar to the role that the relational model plays in the database industry.

Multidimensional data models are designed expressly to support data analyses. A number of such models have been proposed by researchers from academia and industry. In academia, formal mathematical models have been proposed, while the industrial proposals have typically been specified more or less implicitly by the concrete software tools that implement them.

Briefly, multidimensional models categorize data as being either *facts* with associated numerical *measures* or as being *dimensions* that characterize the facts and are mostly textual. For example, a book retailer sells *books* at certain *times* in certain *amounts* and at certain *prices*. A typical fact is a *purchase*. Typical measures are the amount and price of the purchase. Typical dimensions represent the location of the purchase, the purchased book, including its genre and author, and the time of the purchase. Queries then aggregate measure values over ranges of dimension values to produce results such as the total sales per month and author.

Multidimensional data models have three important application areas within data analysis. First, multidimensional models are used in *data warehousing*. Briefly, a data warehouse is a large repository of integrated data obtained from several sources in an enterprise for the specific purpose of data analysis. Typically, multidimensional modeling is applied to such data, as this yields good support for data analyses.

Second, multidimensional models lie at the core of *On-Line Analytical Processing* (OLAP) systems. Such systems provide fast answers to queries that aggregate large amounts of so-called detail data to find overall trends, and they present the results in a multidimensional fashion. Consequently, a multidimensional data organization has proven to be particularly well suited for OLAP. The company behind the so-called "BI Verdict" provides an acid test for OLAP by defining OLAP as "Fast Analysis of Shared Multidimensional Information" (FASMI) [70]. In this definition, "Fast" refers to the expectation of response times that are within a few seconds—such response times enable interactive, or on-line, data analysis; "Analysis" refers to the need for easy-to-use support for business logic and statistical analyses; "Shared" suggests a need for security mechanisms and concurrency control for multiple users; "Multidimensional" refers to the expectation that a data model with hierarchical dimensions is used; and "Information" suggests that the system must be able to manage all the required data as well as derived information.

Third, *data mining*, where the aim is to (semi-)automatically discover hitherto unknown knowledge in large databases, is increasingly being applied to multidimensional data. Indeed, it turns out that a multidimensional organization of data is also particularly well suited for the queries posed by data mining tools.

In summary, the multidimensional data model is very powerful for data analysis. It is widely used in industry, and (at least basic) knowledge of it is needed if one wants to analyze and understand large amounts of data.

1.2   MULTIDIMENSIONAL HISTORY

Multidimensional databases do not have their origin in database technology, but they stem from multidimensional matrix algebra, which has been used for (manual) data analyses since the late 19th century.

During the late 1960s, two companies, IRI and Comshare, independently began the development of systems that later turned into multidimensional database systems. The IRI Express tool became very popular in the marketing analysis area in the late 1970s and early 1980s; it later turned into a market-leading OLAP tool and was acquired by Oracle. Concurrently, the Comshare system developed into System W, which was used heavily for financial planning, analysis, and reporting during the 1980s.

In 1991, Arbor was formed with the specific purpose of creating "a multiuser, multidimensional database server," which resulted in the Essbase system. Arbor, now Hyperion, later licensed a basic version of Essbase to IBM for integration into DB2. It was Arbor and Codd that in 1993 coined the term OLAP [13].

Another significant development in the early 1990s was the advent of large *data warehouses* [48], which were typically based on relational *star* or *snowflake* schemas (see Section 2.5), an approach to implementing multidimensional databases using relational database technology.

In 1998, Microsoft first shipped its MS OLAP Server, which was the first multidimensional system aimed at the mass market. This led to the current situation where multidimensional systems are increasingly becoming commodity products that are shipped at no extra cost together with leading relational database systems.

A more in-depth coverage of the history of multidimensional databases is available in the literature [88].

1.3   RELATED TERMINOLOGY

It is useful to be familiar with a few special terms when studying the literature on multidimensional databases.

OLAP: OLAP abbreviates *On-Line Analytical Processing*. As opposed to the well-known OLTP (On-Line *Transaction* Processing), focus is on data analyses rather than transactions. Furthermore, the analyses occur "On-Line," i.e., fast, "interactive" query response is implied. OLAP systems always employ a multidimensional view of data.

OLAP systems come in three broad categories: systems based on *relational* database management technology, called *ROLAP* systems, systems utilizing non-relational, *multidimensional* array-type technologies, called *MOLAP* systems, and *hybrid* systems that combine these technologies, called *HOLAP* systems.

Data Warehouse: A data warehouse is a repository of integrated enterprise data. A data warehouse is used specifically for decision support, i.e., there is (typically, or ideally) only one data warehouse in an enterprise. A data warehouse typically contains data collected from a large number of sources within, and sometimes also outside, the enterprise.

Data Mart: A data mart is a subset of a data warehouse.

ETL: ETL (Extract-Transform-Load) is the three-step process that feeds a data warehouse. First, an *extraction* process is applied to data from operational source systems, e.g., ERP systems. Second, a *transformation* process is applied in order to change the data format from the source system formats into the data warehouse format. This includes combining data from several different sources and performing *cleansing* to correct errors such as missing or wrong data. Third, a *loading* process is applied to import the data into the data warehouse. ETL is at times also referred to as ETT (Extract-Transform-Transport).

Business Intelligence: Business Intelligence (BI) is the process of making "intelligent" business decisions based on the analysis of available data. From a technology point of view, BI covers the combined areas of data warehousing, reporting, OLAP, data mining, some data visualization, what-if analysis, and special-purpose analytical applications.

1.4   SHORTCOMINGS OF SPREADSHEETS AND RELATIONS

To understand the strengths of multidimensional databases, it is instructive to consider the alternatives. Assume we want to analyze data about sales of books, for which we capture the number of books sold, the book sold, and the city in which it was sold. A simple example with two dimensions is shown in Figure 1.1. When considering how to analyze such data, *spreadsheets* immediately come to mind as a possibility—Figure 1.1 is just a (two-dimensional) spreadsheet.

		City			
		Arlington	Boston	Miami	Springfield
Book	Arlington Road Atlas	2072	2	1	5
	Gone With the Wind	3984	43765	21923	48959
	Italian Cooking	530	9834	4980	7825
	Tropical Food	43	1267	15784	248
	Winnie the Pooh	863	2345	89	1693

Figure 1.1: Spreadsheet with sales data

Our first analysis requirement is that we do not just want to see sales by book and city combined but also the two kinds of subtotals, sales by book and sales by city, as well as the grand total of sales. This means that formulas for producing the (sub)totals must be added to the spreadsheet, each requiring some consideration. It is possible, if rather cumbersome, to add new data to the spreadsheet, e.g., if new products are sold. Thus, for two dimensions, we can perhaps somehow manage with spreadsheets.

However, if we go to three dimensions, e.g., to include time, we have to consider carefully what to do. The obvious solution is to use separate worksheets to handle the extra dimension, with one worksheet for each dimension value. This will work only when the third dimension has few dimension values, and this will only work to some extent. Analyses involving several values of the third dimension are cumbersome, and with many thousands of, say, time dimension values, the solution becomes infeasible. The situation becomes even worse if we need to support four or more dimensions, which in any case will require a very complex set-up.

Another problem arises if we want to group, e.g., the books, into higher-level categories like "Cooking" and "Fiction." Then we must duplicate the grouping information across all worksheets, which results in a system that uses considerable extra space and is very cumbersome to maintain. The essence of the problem is that spreadsheets tie the data storage too tightly to the data presentation—the *structure* of the data and the *desired views* of the data are not separated. However, spreadsheets are good for *viewing and querying* multidimensional data, e.g., using *pivot tables*.

A pivot table is a 2-dimensional table of data with associated subtotals and totals. For example, if we add subtotals by City and Book and a City/Book grand total to Figure 1.1, we have an example of a pivot table as shown in Figure 1.2. To support viewing of more complex data, several dimensions may be nested on the *x* or *y* axes, and data may be displayed on multiple pages, e.g., one for each book. Pivot tables generally also offer support for interactively selecting subsets of the data and changing the displayed level of detail.

		City				
		Arlington	Boston	Miami	Springfield	Total
Book	Arlington Road Atlas	2072	2	1	5	2080
	Gone With the Wind	3984	43765	21923	48959	118631
	Italian Cooking	530	9834	4980	7825	23169
	Tropical Food	43	1267	15784	248	17342
	Winnie the Pooh	863	2345	89	1693	4990
	Total	7492	57213	42777	58730	166212

Figure 1.2: Pivot table with sales data

With spreadsheets falling short in meeting our requirements for the management of multidimensional data, we may then consider using an SQL-based, relational system for data management, as the relational model offers considerable flexibility in the modeling and querying of data. The problem here is that many desirable computations, including cumulative aggregates (sales in year to date), totals and subtotals, and rankings (top 10 selling books), are hard or impossible to formulate in standard SQL.

The main underlying issue is that *interrow* computations are difficult to express in SQL—only *intercolumn* computations are easy to specify. Additionally, transpositions of rows and columns are not easily possible, but rather require the manual specification and combination of multiple views. Although extensions of SQL, such as the *data cube operator* [28] and *query windows* [21], advanced by standards bodies, will remedy some of the problems (see Section 2.8.6 for details), the concept of hierarchical dimensions remains to be handled satisfactorily.

To summarize, neither spreadsheets nor relational databases fully support the requirements posed by advanced data analyses, although it should be noted that these technologies may be adequate under more restricted circumstances. For example, if we have only few dimensions, do not need hierarchical dimensions, and the data volume is small, spreadsheets may provide adequate support. However, the only robust solution to the above problems is to provide data models and database technology that offer inherent support for the full range of multidimensional concepts.

1.5   OLAP VERSUS OLTP

On-Line Analytical Processing (OLAP) is very different from traditional On-Line Transaction Processing (OLTP). When doing OLTP, the focus is on single transactions. Consider, for example, a simple OLTP system used by a book retailer to keep track of the books in stock. When a copy of a certain book B is sold, the system decreases the number showing how many copies of B are in stock. Similarly, the number is increased when more copies are delivered from the publisher. The system only keeps track of the most current data, i.e., what is currently in stock. Detailed information about individual sales is either not stored at all or it is not stored in a format amenable to analysis. The retailer may know who the customer is (for example, if the customer pre-ordered the book), but this information is not kept for the purpose of further analysis. The system stores its information in a relational database that is *normalized* to avoid update anomalies. The OLTP system is thus optimized for retrieving and/or updating a small amount of tuples (e.g., "decrease number in stock of book B by 1").

In contrast, an OLAP system for the book retailer keeps track of each individual sale. This also includes specific information about the sale such as the city where the sale took place, the date of the sale, the salesperson who sold the book, and information about the book. A multidimensional database is used for the storage of such data. This makes it easy to find, e.g., the most sold cookbook and the average number of children's books sold before a holiday. The OLAP system generally never deletes nor updates its data; only additions of new data take place periodically. The OLAP system is thus optimized for retrieving (and summarizing) very large amounts of data (e.g., "find the average monthly number of sold children's books during the previous three years").

CHAPTER 2

Fundamental Concepts

In this chapter, we cover the fundamental concepts of multidimensional databases. We first cover *cubes*, *dimensions*, *facts*, and *measures*. Then we show how these can be represented in relational databases. After that, we present *data warehouses* and *data marts*. We finally cover data analysis and querying using multidimensional databases.

Throughout the book, we illustrate concepts by considering sales in a fictitious book retailer which has shops in different cities. We keep the examples relatively small, but at the end of this and the following chapter, we summarize important concepts by means of more complete and realistic examples.

2.1 CUBES

A *cube* is a truly multidimensional data structure for capturing and analyzing data. A cube generalizes the tabular spreadsheet such that there can be any number of dimensions (and not only two as in spreadsheets). In addition, hierarchies in dimensions and formulas are first-class, built-in concepts, meaning that these are supported without duplicating their definitions. Although the term "cube" suggests that there are three dimensions, a cube can have any number of dimensions. For that reason, the term *hypercube* is sometimes used instead of cube. A collection of related cubes is commonly referred to as a *multidimensional database* or a *multidimensional data warehouse*.

Figure 2.1 illustrates a three-dimensional cube showing sales of books. We assume that Table 1.1 contains aggregated sales data for year 2008. The cube then adds a time dimension and contains sales counts for two cities, two books, and two years. (A real-world cube would contain counts for many more books, cities, etc., but to maintain readability, we keep the numbers low in this example.)

A cube consists of uniquely identifiable *cells* at each of the dimensions' intersections. A non-empty cell is called a *fact*. The example in Figure 2.1 has a fact for each combination of time, book, and city where at least one sale was made. A fact has a number of *measures* associated with it. These are numerical values that "live" within the cells. In the shown example, we have only one measure, the sales count. A cell may also be empty, meaning that there is no information to record for the given dimension values. An empty cell in the cube in Figure 2.1 thus means that the given book was not sold in the given city in the given year. Depending on the specific application, a highly varying percentage of the cells in a cube is non-empty, meaning that cubes range from being *sparse* to being *dense*. Cubes tend to become increasingly sparse with increasing dimensionality and with increasingly finer granularities of the dimension values.

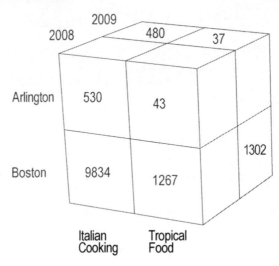

Figure 2.1: Sales data cube

It turns out that most real-world cubes have 4–12 dimensions [48; 88]. Although there is no theoretical limit to the number of dimensions, some tools may experience performance problems when the number of dimensions is greater than 10–15. Generally, only 2 or 3 of the dimensions may be viewed at the same time. However, for low-cardinality dimensions, up to 4 dimensions can be shown by nesting one dimension within another on the axes. Thus, the dimensionality of a cube is reduced at query time by *projecting* it down to 2 or 3 dimensions and *aggregating* of the measure values across the projected-out dimensions. For example, if we want to view sales by City and Time, we aggregate over the entire dimension that characterizes the sales by Book for each combination of City and Time. For the cube in Figure 2.1, we find that the total sales for Arlington in 2008 is 573 by adding up the two numbers 530 and 47.

An important goal of multidimensional modeling is to "provide as much context as possible for the facts" [48]. The concept of *dimension* is the central means of providing this context. One consequence of this is a different view on *data redundancy* than in relational databases. In multidimensional databases, controlled redundancy is generally considered appropriate, as long as it considerably increases the information value of the data. One reason to allow redundancy is that multidimensional databases are often *derived* from other data sources, e.g., data from a transactional relational system, rather than being "born" as multidimensional data, meaning that updates are rare and can be handled more easily [48]. However, there is usually no redundancy in the facts, only in the dimensions.

Having introduced the cube, we cover its principal elements—dimensions, facts, and measures—in more detail.

2.2 DIMENSIONS

The notion of a dimension is an essential and distinguishing concept for multidimensional databases. Dimensions are used for two purposes: the *selection* of data and the *grouping* of data at a desired level of detail.

A dimension is organized into a containment-like hierarchy composed of a number of *levels*, each of which represents a level of detail that is of interest to analyses to be performed. The instances of the dimension are typically called *dimension values* or *dimension members*. Each such value belongs to a particular level.

Figure 2.2 shows the schema and instances of a sample *Location* dimension for the data in Table 1.1.

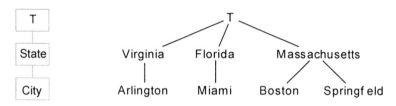

Figure 2.2: Schema and instance for the Location dimension

The Location dimension has three levels, the *City* level being the lowest. *City* level values are grouped into *State* level values, i.e., states. For example, Miami is in Florida. The ⊤ ("top") level represents *all* of the dimension, i.e., every dimension value is part of the ⊤ value.

When using a cube, relevant parts of the data are selected by means of the dimensions. Recall that each cell in a cube is identified by a combination of dimension values. For example, finding the number of sold copies of "Winnie the Pooh" in Boston is just a matter of constraining the Location dimension to "Boston" and the Book dimension to "Winnie the Pooh." It is also possible to group data by means of a dimension. For example, the book retailer could group the data according to the State level in the Location dimension such that the sales counts from a state's individual cities are aggregated into one number for each state. These and other query operations are described in detail in Section 2.8.

In some multidimensional models, a level may have a number of *level properties* associated with it that are used to hold simple, non-hierarchical information. For example, population size of city can be a level property in the lowest level of the Location dimension.

A dimension hierarchy is defined in the metadata of a cube, or the metadata of the multidimensional database, if dimensions can be shared. This means that the problem of duplicate hierarchy definitions as discussed in Section 1.4 is avoided.

Unlike the linear spaces used in matrix algebra, there is typically no ordering or distance metric on the dimension values in multidimensional models. Rather, the only ordering is the containment of lower-level values in higher-level values. However, for some dimensions, e.g., the Time dimension, an ordering of the dimension values is available and can be used for calculating cumulative information such as "total sales in year to date."

In some cases, it is advantageous for a dimension to have *multiple hierarchies* defined on it. For example, a Time dimension may have hierarchies for both *Fiscal Year* and *Calendar Year*. Multiple hierarchies share one or more common lowest level(s), e.g., Day and Month, and then group these into multiple levels higher up, e.g., Fiscal Quarter and Calendar Quarter to allow for easy reference to several ways of grouping. This is illustrated in Figure 3.13. Most multidimensional models allow multiple hierarchies. We will cover multiple hierarchies in Section 3.3.5.

Most models require their dimension hierarchies to form *balanced trees*. This means that a dimension hierarchy must have uniform height everywhere. For example, this means that if some shops are contained in departments, all shops must be contained in departments, even small ones. Additionally, direct links between dimension values can only go between immediate parent-child levels—they cannot jump two or more levels. For the schema in Figure 2.2, this means that all cities have to belong to states. This is problematic when representing Washington, D.C., which belongs to no state. Finally, each non-top value has precisely one parent. Thus, a product must belong to exactly one product group. In Section 3.3, we discuss the relaxation of these restrictions.

2.3   FACTS

Facts are the objects that represent the *subjects* of the desired analyses, i.e., the interesting "things," or events or processes, that are to be analyzed to better understand their behavior. In our book retailer example, we wish to analyze sales. In most multidimensional data models, the facts are *implicitly* defined by their combination of dimension values. If a non-empty cell exists for a particular combination, a fact exists; otherwise, no fact exists for that combination.

Next, most multidimensional models require that each fact be mapped to precisely one dimension value at the lowest level in each dimension. A fact then has a certain *granularity*, which is determined by the levels from which its dimension values are drawn. For example, the fact granularity in our example cube is "Year by Book by City." Granularities consisting of higher-level or lower-level dimension levels than a given granularity, e.g., "Year by Book Genre by City" or "Day by Book by City" in our example, are said to be *coarser* or *finer* than the given granularity, respectively.

It is commonplace to distinguish between two kinds of facts: *event facts* and *snapshot facts*. Event facts (at least at the finest granularity) typically model *events in the real world*. This means that there is exactly one fact for each unique event in the overall real-world process that is captured. With a fine granularity, there would thus be one fact for each particular sale of a certain book. With a coarser granularity there would be one fact for each day when one or more sales of the book occured. Examples of event facts include sales, clicks on web pages, and flow of goods in and out of (real) warehouses.

Facts usually have associated numerical properties that are called measures. These are the properties that the user wants to analyze, e.g., item count and dollar amount. They are described in the next section.

If event facts do not have any measures, they are often called *factless facts*. However, this is an unfortunate term, and we prefer to rather call them what they really are: *measureless facts*. These facts are used to record that some event took place, but here the event does not have a numerical property that describes it further. This could, for example, happen if a company wants to record that it contacts a certain customer at a given time as part of a given marketing campaign.

A snapshot fact, also called a *state fact*, models the *state* of a given process at a given point in time. Typical examples of snapshot facts include the inventory levels in stores and warehouses, and the number of users using a web site. For snapshot facts, the same object (e.g., a specific can of beans on a shelf) with which the captured real-world process (e.g., inventory management) is concerned may occur in facts at different time points.

It is important to understand when event and snapshot facts come into existence. Event facts are made to represent events, e.g., sales, that have happened in the real world. In principle, these events can happen unpredictably at any time. Snapshot facts, on the other hand, are typically made to represent the states at predefined times, e.g., the inventory on the first day in each month.

Often, the above different types of facts coexist in a given data warehouse, as they support complementary classes of analyses. Indeed, the same base data, e.g., the movement of goods in a (real) warehouse, may often find its way into three cubes of different types: warehouse flow, warehouse inventory, and warehouse flow in year-to-date.

2.4   MEASURES

A *measure* has two components: a *numerical property* of a fact, e.g., the sales price or profit, and a *formula* (most often a simple aggregation function such as SUM) that can be used to combine several measure values into one. In a multidimensional database, measures generally represent the properties of the chosen facts that the users want to study, e.g., with the purpose of optimizing them.

The designer of a multidimensional database thus defines what a numerical measure represents. For a snapshot fact representing sales on a given day, the measure would typically represent the amount of items sold during that day. But the designer could also define the measure to be *cumulative* (i.e., only growing), meaning that the measure represents the total sales on that day and all days before it.

Measures take on different values for different combinations of dimension values. Indeed, a measure can in mathematical terms be considered a partial function from the Cartesian product of the dimensions in the cube to a set of numbers like $\mathbb{N}$ or $\mathbb{R}$. The property and formula are chosen such that the value of a measure is meaningful for all combinations of aggregation levels (including the "top" level $\top$). The formula is defined in the metadata and is thus not replicated as in the spreadsheet example.

Although most multidimensional data models have measures, some do not. In these, dimension values are also used for computations, thus obviating the need for measures, but at the expense of some user-friendliness [68].

It is important to distinguish among three classes of measures, namely *additive*, *semi-additive*, and *non-additive* measures, as these behave quite differently in computations.

Additive measure values can be combined meaningfully along any dimension. For example, it makes sense to add the total sales over Book, Location, and Time, as this causes no overlap among the real-world phenomena that caused the individual values. Additive measures occur for any kind of fact.

Semi-additive measure values cannot be combined along one or more of the dimensions, most often the Time dimension. Semi-additive measures often occur when the fact is of type snapshot. For example, it does not make sense to sum inventory levels across time, as the same inventory item, e.g., a specific book, may be counted several times, but it is meaningful to sum inventory levels across books and stores.

Non-additive measure values cannot be combined along any dimension, usually because of the chosen formula. For example, this occurs when averages for lower-level values cannot be combined into averages for higher-level values. Non-additive measures can occur for any kind of fact.

2.5   RELATIONAL REPRESENTATIONS

After having introduced the fundamental concepts in the previous sections, we are now ready to consider how to represent these in a relational database. Note that other representations of multi-dimensional concepts than the relational also exist. Chapter 4 offers further detail. The relational representations are considered here because they are widely used and are easy to grasp. There are two principal ways of representing dimensions in a relational database; we describe the two in turn.

2.5.1   STAR SCHEMAS

A *star schema* has one *dimension table* for each dimension. This table has a key column and one column for each level of the dimension (except $\top$). A column for a level holds textual descriptions of the dimension values at that level. Finally, the dimension table contains one column for each level property in the dimension.

Further, a star schema has a *fact table* that holds a row for each multidimensional fact. The fact table has one column for each measure. In a row, this column contains the measure value for the fact the row represents. A fact table also has one column for each dimension. In a row, these columns contain foreign key values that reference primary key values of dimension tables.

An example star schema instance for the Sales cube is shown in Figure 2.3 (primary keys are underlined). The name "star schema" refers to the observation that if the tables are drawn such that the fact table is at the center and the dimension tables around it, the resulting figure looks (somewhat) like a star. To help visualize this star metaphor, a line is drawn between the fact table and each dimension table in Figure 2.3.

A row in the fact table of the star schema shown in Figure 2.3 holds the sales count for one particular combination of "Book," "City," and "Day." The fact table also has a foreign key column for each of the three dimensions, Book, Location, and Time. The dimension tables have corresponding

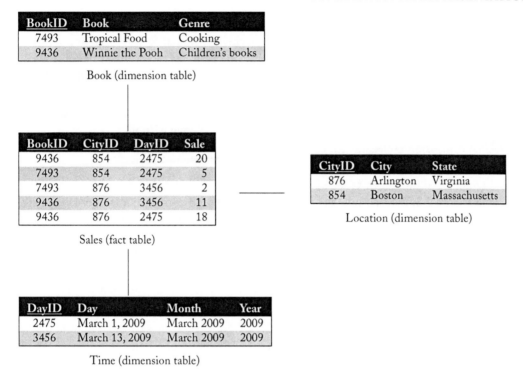

BookID	Book	Genre
7493	Tropical Food	Cooking
9436	Winnie the Pooh	Children's books

Book (dimension table)

BookID	CityID	DayID	Sale
9436	854	2475	20
7493	854	2475	5
7493	876	3456	2
9436	876	3456	11
9436	876	2475	18

Sales (fact table)

CityID	City	State
876	Arlington	Virginia
854	Boston	Massachusetts

Location (dimension table)

DayID	Day	Month	Year
2475	March 1, 2009	March 2009	2009
3456	March 13, 2009	March 2009	2009

Time (dimension table)

Figure 2.3: Star schema for Sales cube

key columns, e.g., "CityID," and one column for each of their levels, e.g., "City," and "State" for the table representing the Location dimension. No column is needed for the ⊤ level, as that column would always hold the same value. The key column in a dimension table is typically a "dumb" integer key without any semantics, i.e., a *surrogate key*. This has several advantages over the option of using information-bearing keys from the source systems, including better storage use, prevention of problems associated with key re-use, better support for dimension updates [48], and more efficient query processing.

It can be seen that there is redundancy in higher-level data. For example, "March 2009" will be present for each day during that month, meaning that the Month column takes the value "March 2009" in 31 rows (assuming that at least one book is sold during each day of that month such that all its days must be represented). However, as dimensions typically take up only 1–5% of the total storage required for a star schema, redundancy is not a problem space-wise; it does not cause update-performance problems. In addition, the updates of dimensions are handled centrally, so it is also possible to ensure consistency. Thus, it is often a good idea to use such redundant dimension tables because these support simpler formulation of (and better-performing) queries than do their normalized counterparts.

2.5.2   SNOWFLAKE SCHEMAS

A *snowflake schema* has a fact table, as does a star schema. Snowflake schemas, however, contain several dimension tables for each dimension, namely one table for each (non-⊤) level. This means that redundancy is avoided, which may be advantageous in some situations, and it renders the hierarchies in dimensions explicit. The dimension tables contain a key, a column holding textual descriptions of the level values, and possibly columns for level properties. Tables for lower levels also contain a foreign key to the containing level. If the tables are drawn with the fact table in the middle and related dimension tables next to each other, the figure looks similar to a snowflake. Thus the name "snowflake schema."

Figure 2.4 shows an instance of a snowflake schema that holds the same data as the star schema in Figure 2.3. For example, the Day table in Figure 2.4 contains an integer key, a date, and a foreign key to the Month table. Note that with this schema, month values will not be replicated. It is, however, harder to query the schema since several joins must be applied. The many joins also make it more time-consuming for the database management system (DBMS) to compute the result of the query.

The choice of whether to use a star schema or a snowflake schema depends highly on the desired properties of the system being developed. Indeed, it is possible to apply snowflaking only partially to a star schema, either fully to only selected dimensions or partially within a dimension. The term *starflake schema* has been applied to the resulting kind of schema. For brevity, we omit a full discussion of this aspect.

2.6   DATA WAREHOUSES AND DATA MARTS

We previously called a collection of related cubes a *data warehouse*. However, there is more than that to say about data warehouses. Bill Inmon, one of the pioneers of data warehousing, defines a data warehouse as a *subject oriented, integrated, time variant, non-volatile collection of data in support of management's decision making process* [35]. In the following, we consider the implications of the different elements of this definition, this way presenting important perspectives on the concept of a data warehouse.

Starting from the back, it can be seen that the purpose of a data warehouse is to support decision making. This is very different from the purpose of an organization's *operational* systems that support the *daily* business processes (handling of orders, deliveries, bills, etc.) but not strategic analysis and decision making. This different focus impacts both *what* data to store in the data warehouse and *how* to store it.

In Inmon's definition, a data warehouse is *subject oriented*. In other words, a data warehouse is designed around the important subjects that concern the business, to allow easy analysis of them. For a book retailer, this includes "sales." A book retailer's data warehouse is thus designed around sales; and to make analysis of them easy, the sales are described by means of entities such as "books" and "shops." In contrast, operational databases are modeled to support daily operations and are heavily influenced by how the daily processes are carried out (i.e., the workflows) in the business and how

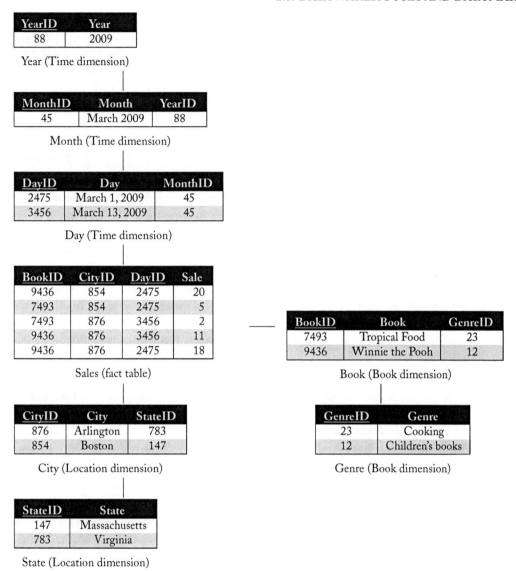

Figure 2.4: Snowflake schema for Sales cube

the operational applications that use them work. The book retailer could, for example, have different operational applications for ordering of books from publishers, distribution to shops, marketing campaigns, and accounting. Analysis of the profit from sales during a campaign then requires the analyst to get data from several different systems that organize their data in different ways and are likely to provide inconsistent data.

Inmon also states that a data warehouse is *integrated*. The data found in a data warehouse will often originate from several different operational systems, e.g., different publishers' book catalogs, a stock keeping system, and the cash registers in shops. Each of these systems may use different formats. For example, one could use the code "P" for paperback, while another system may use "paperback," and a third system may use the numerical code "3." In the data warehouse, a unique code or description is used such that the user does not get confused by different codes that mean the same, or by similar codes with different meanings. We also note that it is considered very bad practice to use cryptic codes in a data warehouse [48]. Instead, understandable, non-abbreviated, textual descriptions should be used (i.e., use "paperback" instead of the confusing "P" or "3"). The different systems might also use different units: for example, some publisher's book catalog may hold the dimensions of books in centimeters while another uses inches. In the data warehouse, only one unit is used. It is the job of the so-called Extract–Transform–Load (ETL) process to extract data from different source systems, clean the data, and transform the data into an integrated format before loading the data into the data warehouse. It is a major part of any data warehouse project to create a working ETL process. We say more about the ETL process in Chapter 4.

A data warehouse is also *time variant*. This means that the data warehouse shows the evolution over time and not just the most recent data, as the operational systems tend to do. In other words, it is possible to see what the modeled world looked like at a certain point in time. For the book retailer, it is then possible to see both how many copies of "Winnie the Pooh" have been sold in total up to now and how many copies were sold by the same time last year. The data warehouse also captures changes in the modeled world, e.g., when one of the book retailer's shops is expanded. It is then possible for the analyst to see the size of the shop at different points in time and investigate how the expansion has affected the sales. We say more about this type of change in the next chapter. We note that virtually any data warehouse has a time dimension.

Finally, Inmon states that a data warehouse is *non volatile*, meaning that neither deletions nor updates are applied to data already in the data warehouse; the only changes are due to the loading of new data. In contrast, the operational systems are designed to support (concurrent) updates and therefore must use advanced transaction handling and normalized designs to enable efficient updates and avoid update anomalies. In a data warehouse, normalization is not needed (recall the star schema from Section 2.5.1). Further, the operational systems typically only include fairly recent data and do not support analysis of historical data.

A *data mart* is generally considered to be a subset of a data warehouse. While a data warehouse is shared throughout an entire organization and can hold data about different subjects, a data mart holds data about a single subject, e.g., sales. Often, a data mart is thus specialized towards the needs of a (sub-)department. So an organization's data warehouse can have cubes and dimensions for both shop sales, Internet sales, stock keeping, and staffing, while a data mart can be built to only hold data about sales, which is the only process the sales department is interested in. Likewise, another data mart can be built to only hold data about staffing, which is what interests the human resource department. Note, however, that the departments should not have their own, private data marts. If two or more departments need to use the sales data, they should both use the same sales data mart. The data warehouse is loaded with data from different operational source systems, while a data mart is loaded with (already cleaned and integrated) data from the data warehouse or even just provides a logical view of data in the data warehouse.

Unfortunately, there is some disagreement in the literature about what the relationship between data warehouses and data marts should be. Some individuals advocate the point of view that a data warehouse consists of a collection of data marts. This view represents a bottom-up approach in which a data warehouse is built by first building the smaller data marts and then merging these to obtain the data warehouse. This is basically the opposite of the definition given above, where the data marts are derived *from* the data warehouse in a top-down manner. Inmon advocates strongly the top-down approach [36]. Among other problems with the bottom-up approach, he points out its lack of the important integration.

An important problem with the top-down approach is the difficulty of the development process—it can be very challenging to build a data warehouse in a top-down manner for a large organization. As a result, the process may be lengthy, with the return on the investment coming late.

In what we may view as an attempt at obtaining the best of both worlds, Kimball, a pioneer of dimensional data warehousing, introduces the concept of so-called *conformed* dimensions and facts. Here, the organization standardizes the facts and dimensions that are to be shared in the organization. This enables a process with both top-down and bottom-up elements, and it renders it possible to combine several marts or cubes in a coherent fashion. The next section covers the modeling process in more detail.

2.7 MULTIDIMENSIONAL MODELING PROCESSES

A key difference of multidimensional modeling from "ordinary" data modeling is that the multidimensional modeler should neither try to include all the available data nor all the existing relationships in the data in the model. Only those aspects that are essential "drivers" of the business should be included.

Another difference is that redundancy is acceptable in a few, well-chosen places (mostly dimensions) if it makes the model more intuitive to the user. For example, we have seen that a month name is repeated in a time dimension for each day of the month. Information about a customer may also be present in both a person-oriented Customer dimension and a group-oriented Demographics dimension.

Kimball [48; 50] organizes the multidimensional modeling process into four subprocesses.

1. Choose the business process(es) to model.

2. Choose the granularity of the business process.

3. Design the dimensions.

4. Choose the measures.

Step 1 relates to the observation that not all business processes are equally important for the business. For example, for a book retailer, there are business processes for *sales* of books to customers and *purchases* of books from publishers, but the sales process is probably the one with the largest potential for increasing the profits, for which reason this business process should thus be prioritized.

Step 2 implies that it is important to use the data granularity that best matches the analysis needs. For example, "individual sales of books" may be captured, or perhaps (the aggregated) "total sales per book per store per day" may be detailed enough, while enabling performance and storage gains. The grain thus decides what a fact means.

Step 3 then goes on to refine the schema of each part of the grain into a complete dimension with levels and attributes. For the example above, a Store, a Book, and a Time dimension are specified.

Finally, the numerical measures to capture for each combination of dimension values, for example dollar sales, unit sales, dollar cost and profit, are chosen in Step 4.

When performing multidimensional modeling "in the large," with many cubes and several user groups, it is very important to ensure compatibility between cubes such that analysis results are comparable. This is ensured by using so-called *conformed* dimensions and measures [48; 50]. Two dimensions are conformed if they are identical or if one is a subset, with respect to dimension values as well as attributes, of the other. Two measures are conformed if they are equivalently defined.

When conformed dimensions and facts are used, it is possible to meaningfully combine data from different cubes. Instead of re-defining the same concept such as "customer" each time it is used, it is far better to have a single definition that is used in all cubes that have a Customer dimension. Likewise, a measure such as "dollar sales" should be defined in the same way (e.g., include sales tax and not be rounded off) such that it, e.g., is possible to compare the planned dollar sales from the Budget cube with the actual dollar sales from the Sales cube.

The creation of conformed dimensions and measures sounds easier than it is, since it often involves interactions, and possibly mediation and negotiations, with different parts of an organization to define, for example, common Product or Customer dimensions that are acceptable to everyone. Large organizations may have 20+ source systems that are used by different departments that all define the concept "customer" in their own way. To obtain a single definition thus requires strong sponsorship and commitment from the management.

2.8   ANALYSIS AND QUERYING

The objective of a multidimensional database is to support data analyses of large amounts of data. Here, we describe the most important operations available to the analyst.

2.8.1   ROLL-UP, DRILL-DOWN, AND DRILL-OUT

Consider the cube shown in Figure 2.5. This cube has the three dimensions Book, Location and Time. Recall that a cube has a cell for each combination of dimension values. Figure 2.5 shows the cells at the intersections of dimension values from the lowest levels in the hierarchies. Note that for practical reasons, we only show very few values in each dimension. A real-world cube would

Figure 2.5: A cube

contain many more dimension values. The analyst would, however, rarely or never consider all the data at the lowest levels in the hierarchies. An analyst may want a less detailed view, for example showing the sales per genre instead of the sales per each individual book. To do this, the analyst can *roll-up* to the Genre level in the Book dimension. By doing that, the analyst considers the cells at the intersections of dimension values from each of the lowest levels of Location and Time and dimension values from Book's Genre level. This is shown in Figure 2.6. The cells in this figure holds the number of sales per genre. Each such number is the sum of sales per book in the genre. Recall that a measure consists of a numerical property and a formula—in this case SUM—that is capable of combining, or aggregating, multiple measure values into one. The number of cookbooks sold in Arlington is given by 2 + 1 = 3 as two copies of "Italian Cooking" and one copy of "Tropical Food" were sold there. To roll-up thus means to go from some level in a hierarchy to a higher level in the same hierarchy in the cube considered. In other words, less detail is seen after a roll-up.

It is also possible to roll-up to the Book dimension's ⊤ level. The ⊤ level, by definition, has only one value. The analyst then considers the cells at the intersections of dimension values from Time and Location and the single dimension value from Book's ⊤ level. This corresponds to the analyst considering the two-dimensional cube (or table) showing the sales per combination of Location and Time values such that the Book dimension is projected out. This is shown in Figure 2.7.

The analyst can also roll-up along any of the other dimensions. If a roll-up is done to the ⊤ level for each dimension, the result is a single cell holding the total number of all books sold.

The opposite operation of the roll-up is called a *drill-down*. When doing a drill-down, more details are shown. It is, for example, possible to drill-down from Book's Genre level to its lowest level. This corresponds to going from Figure 2.6 to Figure 2.5. An analyst typically starts out by

Figure 2.6: The cube from Figure 2.5 rolled-up to Book's Genre level

considering higher-level data and then performs drill-downs to understand the details better. For example, the analyst may consider sales by genre by year and discover that in 2005, many more children's books were sold than in 2004. By doing a drill-down along the Time dimension, the analyst can then discover that the sales were particularly high during July 2005, and by doing a drill-down along the Book dimension, the analyst can see that the increased sales were due to many copies of the, at that time, new book "Harry Potter and the Half-Blood Prince" being sold.

Drilling down offers more detail by "expanding" one or more dimensions from an a non-⊤ level to a lower level. It is also possible to obtain more detail by including an additional dimension (at a non-⊤ level)—we refer to that as a *drill-out*. After a drill-out, the measure values are "spread out" among more cells. Assume, for example, that the analyst is considering the number of books sold in each genre such that only the Book dimension is used in the analysis, as shown in Figure 2.8. The analyst can then drill-out using the Time dimension to see the sales by genre by year. This is shown in Figure 2.9.

2.8.2   SLICING AND DICING

Think of a cube as an onion. When an onion is being prepared for cooking, it may be sliced and even diced into small pieces. In context of a cube, the process referred to as slicing and dicing has a similar effect. When analyzing a cube, the analyst may wish to only consider a subset of the cube. To do this, the analyst can select a specific value for a dimension (e.g., Year 2009 from the Time dimension). This way, a *slice* of the cube is obtained as illustrated in Figure 2.10. It is possible to slice the resulting slice further such that the analyst considers a *dice*. A dice can also be obtained by giving more complex conditions to select a subset of the data (e.g., consider the years 2008 and 2009 and books written by Jane Austen). We call the process of creating these subsets *slicing and dicing*.

	May 1st 2009	May 2nd 2009	May 3rd 2009
Arlington	8	6	9
Boston	13	10	8
Miami	6	3	4
Springfield	13	10	9

Figure 2.7: The cube from Figure 2.6 rolled-up to Book's ⊤ level

2.8.3 DRILL-ACROSS

So far, we have only considered operations on a single cube. Now assume that our book retailer starts an Internet shop to supplement the physical shops. The book retailer does not want to use the existing Sales cube to capture sales from both physical shops and the Internet shop as the Location dimension describes the locations of physical shops in the U.S. The Internet shop may, however, ship to identifiable customers in different parts of the World. The book retailer instead extends the multidimensional database with an Internet Sales cube. Internet Sales and the old Sales cube share the conformed dimensions Book and Time. In addition, Internet Sales uses a Customer dimension.

Children's Books	34988
Cooking	23205
Fiction	50433

Figure 2.8: Sales by Genre

	2008	2009
Children's Books	16904	18084
Cooking	12101	11104
Fiction	26200	24233

Figure 2.9: Sales by Genre by Year

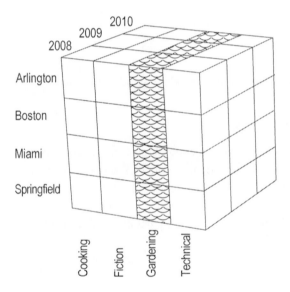

Figure 2.10: A slice

When an analyst considers the sales of books, there are now two cubes to use. However, the analyst wants to see sales figures for the two books that sell the best in total, i.e., considering both shop and Internet sales. To find the best selling books, it is not enough to use the Sales cube to find the books that sell the best in shops and then use the Internet Sales cube to find the books that sell the best on the Internet. To see this, assume that in the physical shops, "Book A" was sold 10,000 times, "Book B" 9,000 times, and "Book C" 8,000 times. From the Internet shop, "Book D" was sold 5,000 times, "Book E" 5,000 times, and "Book C" 4,000 times. If the analyst just found the, say, two best selling books from the cube, the analyst would wrongly conclude that "Book A" sells the best.

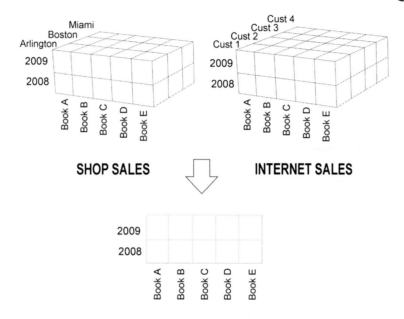

Figure 2.11: A drill-across

To make it possible to see that "Book C" in fact sells the best, the data in the two cubes must be combined.

Data in two cubes can be combined by means of a *drill-across* operation that combines data in two cubes by means of one or more shared dimensions. In relational terms, a drill-across corresponds to a full outer join. Figure 2.11 illustrates the drill-across of the cubes Shop Sales and Internet Sales (for readability, the measure values are not shown).

The resulting cube has as dimensions those dimensions that are shared between the original cubes. Non-shared dimensions can be thought of as rolled up to their top levels in the original cubes before the cubes are combined. The resulting cube inherits all the measures of both the original cubes. These measures are applied to the data in the original cubes. In our example, the analyst should thus consider a *derived measure*, namely the sum of the two cubes' sales measures, to find the best selling books.

When doing a drill-across, the shared dimensions must be conformed, but they need not be identical (remember that two dimensions are conformed if they are identical or if one is a subset of the other). Thus, the cubes may have different granularity. The dimensions in the resulting cube will then have the levels that the original cubes share. For example, the Budget cube can be by genre by year while the Sales cube has finer granularities for the Book and Time dimensions. The data in the Sales cube can then be rolled-up to the Year and Genre levels before the cubes are combined in a drill-across. The measures, showing planned sales and actual sales, respectively, can then be compared meaningfully.

2.8.4 PIVOT TABLES

In Section 1.4, we said that a pivot table is a 2-dimensional table of data with associated subtotals and totals. As an example, consider the pivot table in Figure 2.12 that shows book genres on the y axis and states on the x axis. Note how subtotals and a grand total are also shown.

	Florida	Massachusetts	Virginia	Total
Children's books	2042	8096	4031	14169
Cooking	20764	19174	573	40511
Fiction	50145	68944	12493	131582
Total	72951	96214	17097	186262

Figure 2.12: A pivot table

When using a pivot table, it is typically possible to select the data shown on an axis. For example, the user can pick the genres Cooking and Fiction and the states Florida, Massachusetts, and Virginia and then only sees the relevant subset (rows in this case) of the data and the subtotals and totals of the subset.

A pivot table may also allow the user to use hierarchies to drill-down or roll-up. For example, the analyst can "expand" the Cooking row in the pivot table in Figure 2.12 to see the data for the individual books in this genre. This is shown in Figure 2.13.

	Florida	Massachusetts	Virginia	Total
Children's books	2042	8096	4031	14169
Cooking	20764	19174	573	40511
Italian Cooking	*4980*	*17659*	*530*	*23169*
Tropical Food	*15784*	*1515*	*43*	*17342*
Fiction	50145	68944	12493	131582
Total	72951	96214	17097	186262

Figure 2.13: The pivot table with the Cooking genre expanded

A pivot table can also nest several dimensions on one axis. For example, the analyst can add Year from the Time dimension to the x axis to the pivot table of Figure 2.12. This is shown in Figure 2.14.

Finally, a pivot table can be *pivoted* (as its name suggests!). When pivoting, the dimensions are "rotated." For example, the x and y axes can be swapped, or the Year can be moved from the x axis (as in in Figure 2.14) to the y axis such that there is a row for each genre/year combination.

	Florida		Massachusetts		Virginia		Total
	2008	2009	2008	2009	2008	2009	
Children's books	1013	1029	3784	4312	1452	2579	14169
Cooking	8734	12030	9102	10072	240	333	40511
Fiction	23458	26687	35497	33447	6872	5621	131582
Total	33205	39746	48383	47831	8564	8533	186262

Figure 2.14: A pivot table with two dimensions on the *x* axis

2.8.5 RANKING

When analyzing data, it is often helpful to be able to *rank* it. For example, the book retailer wants to rank the books according to how they sell such that the best selling book is assigned number 1, the second-best selling book is assigned number 2, and so on. It can be relevant to do a complete ranking of all books, but the analyst may also want to find the "top set," e.g., the top-5 best selling books (this is a so-called *top-k* query). The ranking can be based on the number of copies sold, the (aggregated) profit from the books, or on something else. It can also be relevant to do a ranking within another ranking, for example, to find the 5 best selling books within each of the 3 best selling genres.

The MDX query language [82], covered in the next section, offers a number of functions for ranking and sorting multidimensional data.

2.8.6 MULTIDIMENSIONAL QUERYING IN MDX AND SQL

When it comes to the embedding the functionality considered here into a query language, two directions exist: extensions to SQL and dedicated multidimensional languages.

OLAP SQL extensions were pioneered by Gray et al.'s proposal of the CUBE operator [28]. This operator generalizes GROUP BY, crosstabs, and subtotals using the special "ALL" value that denotes that an aggregation has been performed over all values for one or more attributes, thus generating a subtotal or a grand total. "ALL" is thus similar to the $\top$-level in a dimension with a single attribute.

Consider Figure 2.15, which contains a subset of our book retailer's sales data in a standard relational format. The SQL CUBE query below computes the total sales by city and book, with subtotals for both city and book, and a grand total of all sales. The query uses the syntax defined in the SQL:1999 standard.

```
SELECT City, Book, SUM(Sales) AS Sales
FROM SalesTable
GROUP BY CUBE (City, Book)
```

City	Book	Sales
Boston	Gone With the Wind	43765
Miami	Gone With the Wind	21923
Boston	Winnie the Pooh	2345
Miami	Winnie the Pooh	89

Figure 2.15: Sales Table

The query result is shown in Figure 2.16. Note the ALLs that represent the subtotals and the grand total. In most RDBMSs, the ALL value is implemented as a NULL value (possibly with a special function that makes it possible to distinguish it from "real" NULL values).

City	Book	Sales
Boston	Gone With the Wind	43765
Miami	Gone With the Wind	21923
Boston	Winnie the Pooh	2345
Miami	Winnie the Pooh	89
Boston	ALL	46110
Miami	ALL	22012
ALL	Gone With the Wind	65688
ALL	Winnie the Pooh	2434
ALL	ALL	68122

Figure 2.16: Query Result

Another OLAP SQL extension is *window functions* from the SQL:2003 standard. When an aggregate function such as SUM is used in a conventional GROUP BY query, a single result value is produced for a group of rows. With a window function, each row in each implied partition remains, and the aggregate is computed over a set (a "window") of other rows that are related to each row. Consider again the table in Figure 2.15. If we want to compare the amount of copies of each book sold in each city with the total number of sold books in that city, we can use SUM as a window function, as shown in the SQL query here:

```
SELECT City, Book, Sales, SUM(Sales) as Total
OVER (PARTITION BY City)
FROM SalesTable
```

The result of this query is given in Figure 2.17. We partition by City, but the result contains a row for each book sold in each city. The SUM is calculated over all book sales in a city and then associated with each row for a book sale in the city. Thus, all rows for the same city have the same Total value.

City	Book	Sales	Total
Boston	Gone With the Wind	43765	46110
Boston	Winnie the Pooh	2345	46110
Miami	Gone With the Wind	21923	22012
Miami	Winnie the Pooh	89	22012

Figure 2.17: Query result

It is also possible to use RANK that ranks the rows inside each window as a window function. For example, we can rank the books according to sales for each city with the following query.

```
SELECT City, Book, Sales, SUM(Sales) as Total, RANK()
OVER (PARTITION BY City ORDER BY Sales DESC)
FROM SalesTable
```

By now, the SQL standard has adopted the data cube operators [21; 40] along with a range of OLAP functionality such as ranking, percentiles, and windowing, as well as various mathematical and statistical functions. The major RDBMS engines also implement the data cube operators, along with some of the additional functionality mentioned above. However, the syntax may vary slightly from engine to engine. The *iceberg cube* [6] is a further development of the data cube operator that only returns rows where the measure values exceed some threshold (corresponding to interesting "tip of the iceberg" values).

Among the dedicated multidimensional query languages, the by far most prominent and widely used language is MultiDimensional eXpressions (MDX) [82; 89], which originally was proposed by Microsoft, but is now also used in other OLAP products. Unlike the SQL extensions, MDX statements directly produce pivot tables as results, making the integration with OLAP client tools easy.

In MDX, the dimension values are called *members*, and there is a special *Measures* dimension whose values range over the names of the cube measures. This way, any measure value can be referenced using a combination of members (dimension values), including the measure dimension. Dimensions have one or more hierarchies, each with a number of levels. The top level in each dimension is called "(All)" and contains just one one value, like the $\top$ value described earlier. MDX uses dot notation to refer to members, as in [Time].[Calendar].[Month].[May 2009]. An MDX query uses a FROM clause for specifying the input cube. A SELECT clause enables the specification of so-called *axes*, the most important axes being the *query axes*. These specify the hierarchies from which to retrieve the data for multiple members. A query can have up to 128 query axes although queries typically have only a few. There are five "standard" axes, called COLUMS, ROWS, PAGES, SECTIONS, and CHAPTERS, that correspond to well-known report concepts. Additionally, a query may specify a *slicer axis* in the WHERE clause that specifies the hierarchies from which to retrieve the data for a single member.

As an example, consider a cube, SalesCube, with two dimensions, City and Book, and one measure, Sales, built on the data in Figure 2.15. Each dimension has just one hierarchy. The City hierarchy contains a (bottom) City level containing the city names and an (All) level containing the single value "All Cities." The Book hierarchy contains a (bottom) Book level containing the book names and an (All) level containing the single value "All Cities." The MDX query below shows the sales per city and book, including subtotals per city and book, and a grand total, i.e., it is the MDX version of the previous SQL CUBE query example on page 25.

```
SELECT
    { [City].Members, [City].[All Cities] } ON COLUMNS,
    { [Book].Members, [Book].[All Books] }  ON ROWS
FROM [SalesCube]
WHERE ( [Measures].[Sales] )
```

Note the use of the "{...}" notation to specify a set of members to show on an axis, the use of the ".Members" notation to specify the set of members in a dimension, and the use of the "[All ...]" members for referring to the top of the dimension hierarchies. The WHERE clause specifies that the Sales measure should be displayed. Figure 2.18 shows the MDX query result.

Book	Boston	Miami	All Cities
Gone With the Wind	43765	21923	65688
Winnie the Pooh	2345	89	2434
All Books	46110	22012	68122

Figure 2.18: MDX query result

MDX also offers a large number of functions that can be used for navigating in hierarchies (some of these "understand" time semantics such that it, e.g., is possible to compare the current month's sales to sales in the same month last year), performing calculations, ranking data, and much more.

2.8.7 GRAPHICAL QUERYING AND VISUALISATIONS

The MultiDimensional eXpressions (MDX) query language is very powerful and allows the analyst to do advanced querying. However, the language is complex and only IT specialists are capable of writing MDX queries. The typical business analyst must then either get help from such a specialist to write and execute the query or must use a more user-friendly system. Therefore, many OLAP clients with rich graphical user interfaces (GUIs) exist. The OLAP clients provide the user with easy-to-use components (e.g., tables and graphs) that can be shown on the GUI. The user can then perform OLAP operations such as drill-down, slicing, or pivoting by clicking or using menus. Behind the scenes, OLAP clients typically generate MDX or SQL queries.

A popular means to give an overview of summarized data such as key performance indicators (KPIs) or total sale compared to last year is a so-called *dashboard* (think also about the dashboard in a car, which should be readable and understandable in the short timeframe the driver has to look at it). A dashboard, thus, shows a limited number of components, or pre-defined analysis results that give a high-level overview. A dashboard may support further analyses of the data, e.g., by enabling drill-down.

2.9 SUMMARIZING EXAMPLE

We proceed to summarize some of the concepts introduced in this chapter by extending the book retailer example. We follow the four-step process from Section 2.7.

In the first step, we choose the business process to model. For a book retailer, an obvious candidate is *sales of books* as this process directly brings profit to the book retailer. Other processes may also be so important that they should be modeled. That may include *inventory* to avoid out-of-stock or over-stock items, which lead to loss of profit, and *customer flow* to optimize staff scheduling so that long customer waiting times are avoided at the cash registers. Here we choose to only consider sales of books.

In the second step, we choose the grain of the business process. At the finest possible grain, we can capture each individual sale of a book. Alternatively, we can capture aggregated data such as total sales per book per store per date. The finest granularity gives the largest flexibility for analysis (the aggregated sales can be found from the non-aggregated sales figures), but more storage is needed and performance may become worse. To see this, assume that our book retailer has 200 shops, each of which on average sells 1000 book copies each day. If we capture each individual sale, we would on average get $200 \cdot 1000 = 200,000$ new facts each day. It is, however, not the case that all books are equally popular. Some books, e.g., the Harry Potter books, sell extremely well, while others are nearly never sold. Assume that 250 different books are sold (in a number of copies—4 on average) from each shop on an average day. If we only capture the aggregated sales per book per shop per date, we would only get $200 \cdot 250 = 50,000$ new facts each day. In other words, by aggregating the sales, we get 75% less facts and thus need much less storage and get better performance, as there is less data to consider in queries. With the aggregated data, there are, however, aspects of the business we cannot investigate. For example, we cannot see during which hours books sell the best, and we cannot use *data mining* to investigate whether certain books are likely to be sold together. It is, thus, very important to understand the needs of the business when creating a multidimensional database. In general, the finest possible granularity should be used. For now, however, we assume that the book retailer only needs the aggregated sales per book per shop per date.

In the third step, we design the dimensions. From the grain "sales per *book* per *shop* per *date*," we can see three (emphasized) dimension candidates. In the examples throughout the chapter, we have used a Location dimension with City as the lowest level. However, the book retailer may have several shops in one city. Thus, we now use a Shop dimension instead. We also rename the Time dimension to the Date dimension to avoid confusion about whether the dimension represents dates (e.g., May 26th 2010) or specific times (e.g., 11:02 a.m.). When we pick attributes for the dimensions, as much relevant context as possible should be included. The "relevant" context is what business analysts can make use of. For example, it is relevant to know how large a shop is (a large shop should probably sell more than a small shop), but the yearly consumption of electricity in a shop does not seem relevant. Note that there is also a trade-off between the utility of an attribute and the cost or difficulty of obtaining the necessary data. In this example, we choose to include the following attributes.

Book dimension

- BookID (surrogate key)

- ISBN number (a business key as each book has a unique ISBN number)

- Title

- Edition

- Author

- Original title (the book may be a translated publication)

- Publisher

- Published (when the book was released)

- First published (the book may be a republishing)

- Genre

- Language (the book retailer sells books in several languages)

- Number of pages

- Binding (paperback/hard-cover)

Date dimension

- DateID (surrogate key)

- Date

- Month

- Year

Shop dimension

- ShopID (surrogate key)

- Address

- City

- State

- Size (floor space)

- Shelf space

- Opened (the date when the shop opened)

- Has café

The dimension schemas are shown in Figure 2.19. The attributes not shown here are level properties of the lowest level. The hierarchies for the Date and Shop dimensions are straightforward.

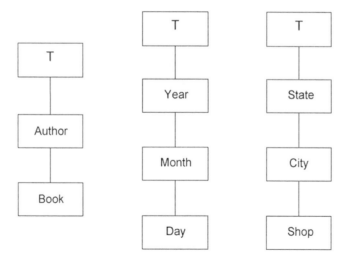

Figure 2.19: Schemas for the dimensions (level properties not shown)

For the Book dimension, we choose to make it possible to roll-up the individual book titles to authors. This will be directly supported by client tools that understand hierarchies (e.g., MDX-based solutions). However, it could be that the analysts need other hierarchies, for example, to roll-up sales based on publisher of the book or genre of the book. We will come back to this in the next chapter.

In the fourth step, we choose measures. An obvious choice is the number of sold books (with the sum formula for aggregation). Also the sales price in dollars for the sold books seems relevant (prices may change over time and location, e.g., due to special campaigns).

This completes the modeling (although a real-world design document would be much more elaborate and precise). We have now modeled a multidimensional database that helps the book retailer's analysts in understanding how books sell. However, our multidimensional model has a number of limitations and inaccuracies. First of all, we only have a single Author attribute. This is fine when a book has a single author. But in the real world, many books are co-authored by several authors. We could then choose to write all their names in the Author attribute or only write the first author's name. However, both of these approaches yield incorrect results when we want to roll-up to the Author level. What we need is a way to represent the many-to-many relationship between books and authors. Regarding the hierarchies, it would also be desirable if we had more hierarchies on the Book dimension. For example, it could make sense to have a hierarchy with the levels Book, Subgenre, Genre, and All, or a hierarchy with the levels Book, Publisher, and All. Another problem is that there is no way to capture changes with the current modeling. A shop may, for example, be expanded such that its floor area increases. If we update the shop's dimension value, old facts will incorrectly be associated with the new size. All of these issues (and many others) are dealt with in the next chapter.

2.10 EXERCISES

1. Illustrate how the data in the cube in Figure 2.20 can be presented in a two-dimensional table by nesting its dimensions. Assume that the cell that cannot be seen in the drawing contains the number 7201.

2. Show how the cube in Figure 2.20 could look if we roll-up to the $\top$ level in the Location dimension.

3. Show how the result from Exercise 2 looks if we drill-down in the Date dimension.

4. Define the schema for a Date dimension with at least 5 levels and identify relevant attributes. Can a Week level be included? Should the attribute Weather be included if the dimension is designed for an ice-cream retailer?

5. For each of the following statements, decide what type (event or state) of fact we are considering.

    (a) A fact exists for (d, p, c, s) if product p was sold to customer c from shop s on day d. The measure is the total price.

    (b) A fact exists for (d, p, s) if product p was sold from shop s on day d. The measure is the total price.

    (c) A fact exists for (d, p, s) for each product p, each shop s, and each day d. The measure is the total inventory of p in s on d.

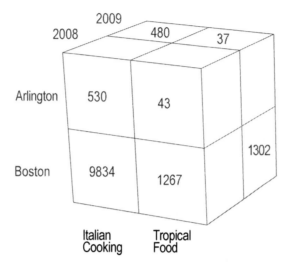

Figure 2.20: A cube

(d) A fact exists for (d, p, s) for each product p, each shop s, and each day d. The measure is the total amount of p sold in s in the year until day d.

(e) A fact exists for (c, e, d) when a call from customer c was answered by employee e on day d. There is no measure.

(f) A fact exists for (c, e, d) when a call from customer c was answered by employee e on day d. The measure is the length of the call in seconds.

6. For each of the following statements, determine whether the measure is additive, semi-additive, or non-additive.

   (a) A fact exists for (d, p, c, s) if product p was sold to customer c from shop s on day d. The measure is the total price.

   (b) A fact exists for (d, p, s) if product p was sold from shop s on day d. The measure is the total price.

   (c) A fact exists for (d, p, s) for each product p, each shop s, and each day d. The measure is the total inventory of p in s on d.

   (d) A fact exists for (d, p, s) for each product p, each shop s, and each day d. The measure is the total amount of p sold in s in the year until day d.

   (e) A fact exists for (b, c, d) if currency c was exchanged in bank branch b on day d. The measure is the total amount exchanged in US Dollars.

(f) A fact exists for (b, c, d) if currency c was exchanged in bank branch b on day d. The measure is the average exchange rate between c and US Dollars for all transactions involving c on day d.

(g) A fact exists for each (b, c, d) where c is a currency, b a bank branch, and d a date. The measure is the total amount of c kept in b on day d.

7. In this exercise, we consider a chain of stores selling newspapers and magazines. The chain sells many different kinds of publications (e.g., relating to fashion, children, cars, sports) from many different publishers. The types of stores range from small corner shops to super stores with co-located cafés. However, the chain is a bit old-fashioned, as each shop manager at the end of each day has to enter information into a spreadsheet about how many copies were sold of each publication that day. The spreadsheet is then sent to the headquarters. Currently, this is the only way headquarters can collect and analyze sales data from the shops. The management now wishes to gain more insight into the sales of the different publications (and types of publications) from each shop. Design a data warehouse that can help the management. Include relevant hierarchies and attributes. Remember that a certain publication has many "editions."

8. Draw a star schema for the data warehouse designed in Exercise 7.

9. Draw a snowflake schema for the data warehouse designed in Exercise 7.

10. Write SQL queries that find the total sales of sports publications in 2009 when using the star schema and the snowflake schema from the two previous exercises. Discuss pros and cons in each case.

CHAPTER 3

Advanced Concepts

Having covered the fundamental concepts of multidimensional databases in the previous chapter, we proceed to consider advanced concepts that provide solutions to the problems pointed out at the end of the previous chapter where many-to-many relationships were not supported and where hierarchies were somewhat limited. First, we look into how to handle changes in dimensions. The examples will extend the simple examples used throughout Chapter 2. The complete example from Section 2.9 is extended in Section 3.4.

3.1 SLOWLY CHANGING DIMENSIONS

Like any database, a multidimensional database or data warehouse models selected aspects of some reality. Which specific aspects to capture depends on the intended uses of the database. Due to the dynamic nature of reality, the modeled reality as well as the uses of the data warehouse change over time. This implies that the database must be able to evolve in order to continue to serve its purposes. In this section, we consider the handling of changes in a database that is represented by means of a star schema. As described in Section 2.5.1, such a database has a central fact table and a number of dimension tables.

3.1.1 THE PROBLEM

Recall that a data warehouse models some real-world process that we are interested in studying. The example we consider focuses on the sales process for books. As the process evolves, new rows are inserted into the fact table and into the dimension tables. In our example in Figure 2.3, when sales come in for a new day, that day is inserted as a row into the Time dimension table, and a row is inserted into the Sales fact table for each combination of book and a city for which there was at least one sale during that day. If a new book starts to be sold, that book is inserted into the Book dimension table, and if sales start to occur in a new city, that city is inserted into the Location dimension table.

This scenario represents the desired evolution of a data warehouse. In practice, however, it is necessary to be able to cope with other types of change, including the so-called *slowly changing dimensions*, which occur when the existing rows in dimension tables need to be updated [48]. The mental image intended by this naming is that although existing rows in dimension tables need to be updated occasionally, this happens infrequently.

Consider the star schema shown in Figure 2.3 with the slightly revised Book dimension table shown in Figure 3.1. The table now includes a Rating column that indicates how well the customers

BookID	Book	Rating	Genre
7493	Tropical Food	4 stars	Children's books
9436	Winnie the Pooh	5 stars	Children's books
9948	Gone With the Wind	4 stars	Fiction
9967	Italian Food	4 stars	Cooking

Book (dimension)

Figure 3.1: Revised Book dimension table for the Sales cube

like the book. Both the rating and the genre of a book may change over time. For example, the rating of "Gone With the Wind" may drop to "3 stars" or the genre of the book "Italian Food" may be refined to "Mediterranean cooking." The fundamental problem with simply updating the rows accordingly is that these rows are already being referred to by fact table rows with old sales. These already existing fact table rows rely on the existing dimension table rows with their specific attribute values. When updating such attribute values, the old fact table rows refer to dimension table rows that have changed. For example, purchases of "Gone With the Wind" when rated as "4 stars" now appear to be purchases of a book with only a "3 stars" rating. This way, incorrect information is created. Next, the fundamental problem with not changing the rows to reflect changes of rating and genre classifications is that the data warehouse is outdated. New rows entered into the fact table really need to refer to the updated Book dimension table rows. As an aside, observe that an update of the genre of the "Tropical Food" row from "Children's books" to "Cooking" is acceptable in that this change is a correction of an error.

3.1.2   SOLUTIONS

We consider three approaches to addressing the changes that may occur in dimension table rows.

The first approach is to simply overwrite the old attribute values. This is called a *type 1 update*. As already pointed out, old fact table rows now refer to dimension table rows that have changed, and if the original state of the data warehouse was correct, the data warehouse now contains incorrect information. The good news is that new fact table rows will be able to refer to correct dimension table rows.

This approach is easy to implement, and if the dimension table updates are simply error corrections, as in the case of "Tropical Food" being classified as a children's book, the solution is ideal. Further, there may be cases where the inaccuracies introduced by the updates are considered as unimportant. For example, this might be the case for the genre refinement. However, the bottom line is that this approach basically ignores the fundamental problem.

The second approach, called a *type 2 update*, is to version the rows in the dimension tables. A change is captured by inserting a new row with the updated attribute values, leaving the existing row unmodified. This has the conceptual effect that the dimension tables go from recording rows to recording "versions of rows." The primary key column thus has to be generalized to capture versions of rows instead of rows. This generalization is straightforward when surrogate keys are used, as is the recommended approach. With this approach, the old fact table rows continue to refer to the original dimension table rows, while new fact table rows refer to the new dimension table rows. In our example, if the rating of "Gone With the Wind" changes, we create a new row for this book that differs from the existing row only on the Rating value, and new sales of "Gone With the Wind" will refer to this new row.

This approach enables the capture of correct information in the data warehouse. It has the effect that the dimension tables increase in size. This is usually not a problem, as the space used by dimensions typically is much smaller than the space used to store facts. Now that the dimension tables change over time, it may be relevant to capture accurately these changes. In case a time dimension is present, which is typically the case in a data warehouse, the first time a fact table row refers to a row in a dimension table can be determined, and this time can serve as a conservative bound on when the dimension table row came into existence. The exact time when a change happens in a dimension can be captured by inserting a special row in the fact table. This row refers to the new dimension row and to the row in the time dimension table that represents the time when the change happened. This is, however, cumbersome to use in queries. A simpler alternative is to introduce two time-valued columns in the dimensions where changes may occur. These columns record the times when the rows became valid and invalid, respectively. Note that for a version that is still valid, we can let the column telling when it becomes invalid point to some date far into the future. A NULL value could also be used, but that renders querying more difficult.

The solution can be extended with a column that for each row tells whether the row is the most recent version of the row. This column is then updated in the old version when a new version is inserted. A column can also be added to record the version number of each row. Note that the two latter columns (recording if a version is the most recent and the version number, respectively) in principle can be deduced from the two time-valued columns (recording when versions become valid or invalid, respectively). However, some queries are much easier to formulate when the two latter columns are present in the dimension table. Figure 3.2 shows the Book dimension table when this scheme is applied.

BookID	Book	Rating	Genre	ValidFrom	ValidTo	Newest	Version
7493	Tropical Food	4 stars	Children's books	2006-03-01	2008-12-31	No	1
9436	Winnie the Pooh	5 stars	Children's books	2000-01-01	9999-12-31	Yes	1
9948	Gone With the Wind	4 stars	Fiction	1999-06-01	2008-10-15	No	1
9967	Italian Food	4 stars	Cooking	2003-04-05	2009-05-01	No	1
9995	Gone With the Wind	3 stars	Fiction	2008-10-16	9999-12-31	Yes	2
10100	Tropical Food	4 stars	Cooking	2009-01-01	9999-12-31	Yes	2
11319	Italian Food	4 stars	Mediterranean cooking	2009-05-02	9999-12-31	Yes	2

Book (dimension)

Figure 3.2: Book dimension table with row versions ("type 2 updates")

The third approach to support slowly changing dimensions is quite different. For each dimension table column that may change, an additional version of that column is introduced. This is called a *type 3 update*. It has the effect that we are able to record two values for columns that may change. We use one to record the current value, and we use the other to record the previously current value. When this scheme is applied to our example, we obtain the dimension table shown in Figure 3.3.

BookID	Book	Rating	OldRating	Genre	OldGenre
7493	Tropical Food	4 stars	4 stars	Cooking	Children's books
9436	Winnie the Pooh	5 stars	5 stars	Children's books	Children's books
9948	Gone With the Wind	3 stars	4 stars	Fiction	Fiction
9967	Italian Food	4 stars	4 stars	Mediterranean cooking	Cooking

Book (dimension)

Figure 3.3: Book Dimension table with versioned columns ("type 3 updates")

With this approach, fact table rows from before and after the most recent change to an attribute refer to the same dimension table row; and as this row holds both the current value and the most recent, previously current value, fact table rows refer directly to two values of the same attribute. This arrangement may be utilized to analyze data across changes. For example, one can study the difference in the distribution of sales across two different genre classifications of books in preparation for deciding whether or not to introduce a new genre classification of books that is more useful for analyses.

This approach is limited by its ability to capture only two values for each attribute that can change, and this renders the approach useful only for certain special cases. It should also be noted that it is not possible to capture when the changes occur for the attributes. The approach may, however, be generalized to support cases where changes happen for all dimension members at given times. Sales districts could, for example, be redefined for each year. In the Shop dimension, there could then be an attribute for each year showing the sales district for that year.

When the type 2 or type 3 approaches are applied, to be able to reflect changes in the real world, it often makes sense to combine them with the type 1 approach for the handling of error corrections. In Figure 3.2, we created a new row version for the book "Tropical Food," which was defined erroneously to be a children's book. Creating a new version only fixes the problem from the point on when the new version took effect; old facts that do not refer to the new version still refer to the wrong information. A better solution for this error is to simply change the wrong genre value for the book.

We have only considered star schemas in this section. In a snowflake schema, slowly changing dimensions may, however, also occur. The type 1 and type 3 approaches can then be used straightforwardly. The type 2 approach can also be used, but to avoid an overly complex scenario, the row versioning should be kept in the dimension table for the lowest level. Thus, if a new version is created at a higher level, corresponding new version(s) should also be created for the lower levels.

We conclude by remarking that the most versatile technique for handling slowly changing dimensions is the row versioning approach ("type 2"), and this approach is generally recommended.

3.2   OTHER SPECIAL KINDS OF DIMENSIONS

3.2.1   MINIDIMENSIONS

In Section 3.1, we considered slowly changing dimensions. It is natural to ask when slowly changing dimensions become rapidly changing, and to consider how to address such cases.

Assume that our book retailer has access to detailed information about customers who have joined a loyalty program. This detailed information includes address, birthday, marital status, number of children, education, job category, yearly salary, etc. When members of the loyalty program purchase products from a shop, they show their membership cards to get rebates. This arrangement allows the book retailer to track the purchases of the members in considerable detail.

The book retailer can then also include information about the individual member in the multidimensional database by means of a Customer dimension. That way, the book retailer can analyze how book sales are related to age, level of education, etc. We assume that the retailer includes 30 attributes describing the customers. However, customer information will change over time, such as when a customer moves or gets a raise, and so the Customer dimension is a slowly changing dimension. The retailer can use type 2 updates where new versions are created to track these changes. A new version is created whenever one or more of the attributes get a new value. Thus, the Customer dimension is going to have many dimension value versions (or row versions in a relational OLAP system).

When changes to a dimension are frequent and we use row versioning, the effect is that the size of the dimension may get very large. This may occur in the example above. Still, with proper indexing, it is generally possible to handle large, rapidly changing dimensions.

In special cases where the versioning simply yields a dimension that is too large, one possible remedy is to split the dimension into two (in relational OLAP, both dimension tables are referenced from the fact table). The idea is to break off the attributes that change often and place them in a new dimension, called a *minidimension*.

The minidimension can hold all possible combinations of the values of the often changing attributes. Here, "all possible combinations" can either mean the Cartesian product of the attribute values, i.e., all combinations possible from a theoretical point of view, or it can mean all combinations that actually occur in real life. The size of the latter set is often only a fraction of the size of the Cartesian product. To reduce the size of the minidimension, the detailed attribute values may be replaced by ranges of values. For example, if a customer dimension originally recorded the average annual income of the customers, the new minidimension may instead record annual income *ranges*, using, e.g., 5 ranges. This limits the number of possible values in the minidimension.

In our example, the retailer can thus create a minidimension, Profile, that describes "customer profiles" (and not the individual customers) and holds all the attributes that change often, such as salary, number of children, and credit rating, while the (relatively) static attributes (e.g., education and birth date) and those that only describe an individual customer, such as address, remain in the Customer dimension.

Since all possible combinations are represented in a minidimension, there is no need for the handling of changes in a minidimension; instead, changes in the modeled world are captured by means of the facts. For example, if a customer gets a higher salary, it is not necessary to update anything in the relevant minidimension. Rather the change is captured the next time a fact occurs for the given customer. The new fact will then reference the minidimension's dimension value that now covers the salary.

The negative consequences of using a minidimension are that the data becomes less detailed and that some queries may become harder to formulate because the attributes they refer to are now in two dimensions rather than in one. In a relational OLAP-based implementation, the fact table needs an extra foreign-key column. Since a data warehouse normally holds a large number of fact rows, the increased space needed is not negligible. Another consequence is that some changes only can be captured when a fact occurs. With a slowly changing dimension, it is possible to capture a change in, e.g., demographic information about a customer's salary, on the day the change happens even if there is no fact for the customer for that day. If the information, on the other hand, is placed in a minidimension, the change cannot be captured until the next time a fact occurs for that customer.

3.2.2   OUTRIGGERS

We have just seen how the book retailer can describe customer profiles with a minidimension, Profile, to keep the Customer dimension small in size even when it changes rapidly. Now assume that the book retailer wants to be able to see the current profile associated with a given customer without relying on the existence of a fact for the customer and the current profile. A naive solution would be to include the 30-some attributes from the Profile dimension in the Customer dimension and then use type 1 updates to represent the current profile. However, it does not seem like a good idea to maintain some 30 Profile attributes in both the Profile minidimension and in the Customer dimension. A better solution is to reference the Profile minidimension directly from the Customer dimension (and use type 1 updates to only reference a customer's current profile). A dimension that is referenced from another dimension is called an *outrigger*.

Outriggers are used in relational OLAP environments where the outrigger's dimension table is referenced by a foreign key in another dimension table. To use the outrigger, a join between the dimension table and the outrigger is performed. Note that the use of an outrigger is different from using a snowflake schema. The outrigger does not have to be normalized.

A dimension can be used both as an ordinary dimension and as an outrigger at the same time. This is, for example, the case for the Profile dimension described above. Another typical example is a time dimension. A time dimension is normally part of a cube such that a fact is related to a certain date, but the time dimension may also be referenced from other dimensions, e.g., from the Customer dimension to keep track of a customer's birthday or from the Book dimension to keep track of when a book was published.

3.2.3 DEGENERATE DIMENSIONS

In the following, we assume that our book retailer now uses a finer granularity and keeps track of each individual transaction where one or more books are sold (i.e., not only the sales per book per day per location). The dimensions Book, Time, and Location are still used, but these are not sufficient as we cannot identify the individual transactions. So we need another dimension to be able to identify a unique transaction. This dimension thus holds a unique identifier from each transaction, e.g., the unique number the cash register prints on the receipt or just a "dumb" integer value that is increased for each transaction. However, there are no other descriptive attributes and the dimension's only levels are the lowest level for the unique identifier and the special ⊤ level for all transactions. Such a dimension that consists only of a single identifier is a *degenerate dimension*.

In a relational representation, as covered in Section 2.5, there is no need to create a separate dimension table for a degenerate dimension. If we did, the fact table would have a foreign key referencing the dimension table, which would only have a single attribute. Instead, it is easier and yields better query performance (as joins can be avoided) to just place the identifier attribute of the degenerate dimension directly in the fact table, thus avoiding the separate dimension and the foreign key. Note that it then is important to use an integer as the identifier to keep the size of the fact table down.

Figure 3.4 shows the fact table for our book retailer when we have added the degenerate dimension Transaction (the fact table is the the only place where the transaction's identifier, TransactionID, is represented). Note that we have added the measure Price, as the finer granularity allows us to keep track of the price a book was sold for in a given transaction (which may differ from transaction to transaction and from shop to shop due to special offers, campaigns, etc.).

BookID	CityID	DayID	TransactionID	Sale	Price
7493	854	2475	102	1	10
7493	854	2475	123	1	10
7493	854	2475	232	2	20
7493	854	2475	244	1	10
7493	876	3456	400	1	15
7493	876	3456	523	1	15
9436	876	3456	523	1	12

Sales (fact table)

Figure 3.4: Revised fact table for the Sales cube

3.2.4 JUNK DIMENSIONS

With the transaction-level granularity introduced and used in the previous section, it is of relevance for our book retailer to provide even more context about each book being sold in a transaction. Now we also want to capture how the book is being displayed (with the three options "Displayed in prime location," "Displayed in secondary location," and "Not displayed"), if the book is discounted (with the two options "Discounted", "Not discounted"), and finally whether some promotion event takes place (with the four options "Release," "Book signing," "Other event," "No event").

To capture these options, we could add the three dimensions Display, Discount, and Event. Each of these dimensions has very few (between two and four!) dimension values. However, as each of the three dimensions has very few dimension values, it is viable to "merge" them into one dimension, Promotion. This dimension holds all possible combinations of the options, in total $3 \cdot 2 \cdot 4 = 24$ dimension values. We call such a dimension with combinations of unrelated values, a *junk dimension*. A junk dimension should only be used to group flags and low-cardinality descriptions as there would otherwise be too many possible combinations.

While the the querying can become slightly harder with a junk dimension instead of several independent dimensions, this arrangement reduces the dimensionality of the cube. In a relational representation, this leads to huge space savings, as the fact table then only needs one extra column to accommodate the new contextual information instead of three extra columns (remember that the fact table is likely to hold many millions of rows).

Kimball & Ross [48] give a good mental image of the junk dimension by comparing it to a kitchen's junk drawer where scissors, rubber bands, tape, etc. are kept. It would be nice to have a drawer for each category, but there is not enough space for that. Therefore, we make a pragmatic solution and put all these unrelated items in one drawer, the junk drawer.

3.2.5   TIME DIMENSIONS

Most real-life multidimensional databases model processes or states that evolve over time (in our example, the book retailer sells books every day). To model this evolution, *time dimensions* are used. Time dimensions are thus found in nearly all multidimensional databases. Recall from Section 2.6 that Inmon views the presence of a time dimension as a defining characteristic of a data warehouse.

There are two kinds of time dimensions: A *date dimension* represents the date (e.g., "November 11, 2009") while a *time-of-day dimension* represents the clock time (e.g., "11:47 a.m.") during some unspecified day. Whether or not both kinds of time dimensions are present in a multidimensional database depends on the intended use of the database. Sometimes it is enough to capture the date and not the more detailed time of day. The granularity of these dimensions may also vary depending on the intended usage. For example, data could be collected at a monthly level instead of a daily level (meaning that the date dimension would represent a specific month, e.g., "November, 2009," instead of a specific day).

It would be possible to include both the date and the time of day in a single dimension. However, this dimension would quickly grow to be huge. Assume that such a dimension represented each minute in each day. For a single day, there would thus be $60 \cdot 24 = 1,440$ dimension values. For a single year, there would be $1,440 \cdot 365 = 525,600$ dimension values. In comparison, there are only 1,440 dimension values in the time of day dimension (independently of how many years are represented in the date dimension) and 365 dimension values per year in the date dimension.

Further, it may be more natural for analysts to consider the time of day and the date independently. For example, this would make it possible to create a pivot table with dates at the month level on one axis and times of day at the hour level along the other axis, thus enabling analysis of how sales during the day are affected by the season.

Even if a DBMS to be used for data warehousing offers a DATE attribute type, the use of time dimensions is recommended instead of just having a simple DATE attribute in the fact table. By having time dimensions, domain-specific knowledge can also be represented and made easily available to analysts. For example, it can be captured easily whether a day is a part of a holiday season or whether a special event like the Soccer World Cup takes place during a particular day. This kind of information cannot be extracted from an ordinary DATE type. Also, the time dimensions are easier to use in queries since no calender logic must be performed in the query when a dimensional, hierarchical representation is in place. Time-related drill-down and roll-up operations are thus also much easier to perform.

To make a time dimension as useful as possible, it is important to include many descriptive attributes—also if these attributes represent the same thing. For example, there is a one-to-one relationship between DayNumberInWeek (with values 1, 2, ..., 7) and WeekDay (with values Monday, Tuesday, ..., Sunday), but it is still useful to include both attributes to make the formulation of queries and browsing of dimension values easy and intuitive. Another example is the number of days in a certain month, which in principle could be derived from the month (and the year, in case of February), but instead of burdening the user with this, an attribute, e.g., called NumberOfDaysInMonth, should be included in the date dimension.

A date dimension typically has more than one hierarchy as both calendar year and fiscal years are often captured. We return to multiple hierarchies in Section 3.3.5.

3.2.6   DATA QUALITY DIMENSIONS

Sometimes it is possible to evaluate the quality of the data in a database. Assume, for example, that one of the book retailer's shops usually sells 3–5 copies of a certain book each day, but one day appears to have sold 1,000 copies of that book. This is likely to be an error, but it could be true due to a discount or a new film being based on the book. Thus, it can be dangerous to just remove the data. Instead, a special dimension can be used to describe each fact. Which dimension values to include in such a *data quality dimension* depends on the business needs. Typical values are "Normal value," "Out-of-bounds value," "Unlikely value," "Verified value," "Unverified value," and "Uncertain value."

Inclusion of all the data even if the associated quality is questionable gives the analysts the full picture. If they wish to only use high-quality data, they can constrain the data quality dimension to "Normal value" or "Verified value." If they wish to see all data (including possibly wrong data), they can ignore the data quality dimension altogether.

3.3   ADVANCED HIERARCHIES

The hierarchies we considered in Chapter 2 were *simple* hierarchies. We assumed that they were *balanced* (i.e., in any given instance of the hierarchy, all leaves belong to the schema's lowest level), *covering* (i.e., in any given instance, each path starts at the root and then goes to the level immediately below in the schema and then to the next level immediately below and so on, without skipping any level), and *strict* (i.e., in any given instance, no dimension value has more than one parent). Intuitively, this means that the instance forms a balanced tree. In this section, we consider more general hierarchies that do not fulfill these requirements, and we discuss how to support such hierarchies. Further, we describe how a dimension can have several hierarchies rather than just one as assumed in Chapter 2.

3.3.1   PARENT-CHILD HIERARCHIES

In the hierarchies we have considered so far, level values of one type have been grouped into level values of containing type. For example, cities were grouped into states. In some cases, the type of the *parent* level is, however, not different from the type of the *child* level. Thus, the parent can itself be a child of another parent of the same type. We represent this by means of a *parent-child hierarchy*. A typical example is an Employee dimension. Employees have managers that are employees themselves and have other managers. For example, assume that the employee Smith has Johnson as manager. Johnson has Davis as manager and Davis has Douglas as manager, etc. Smith, Johnson, Davis, and Douglas are all represented as Employee-level values.

It is not practical to use any fixed number of levels in the Employee dimension to represent such a hierarchy. For the book retailer, a small shop may have only a shop manager and few assistants (i.e., two levels), while a big shop also has mid-level managers (i.e., three or more levels). Instead of having a fixed number of levels, we let a parent-child hierarchy's schema have a single level. However, in instances of the hierarchy, we allow a dimension value (from the single level) to have another dimension value (from the same level) as parent. In that way, we in effect can have an unlimited number of levels in an instance. To roll-up, the measure values of a parent's children are combined into a single value for the parent by means of the measure's formula.

In relational OLAP, a parent-child dimension can be implemented by having a single dimension table with a foreign key column that references the table's primary key column. Consider as an example the Employee dimension in Figure 3.5 and its relational representation where ManagerID is the foreign key column referencing EmployeeID.

In some systems, it is possible to assign names to the levels in a parent-child hierarchy. We can choose to name the top-level "President," the next level "Vice-president," the level below that "Shop Manager,", and the following level "Assistant." In our example, that means that Dougles belongs to the President level, Davis and Button belong to the Vice-president level, Johnson and Brown belong to the Shop Manager level, and Smith, Geller, and Hansen belong to the Assistant level.

The above method is capable of capturing a parent-child hierarchy, and some relational OLAP systems are capable of using such a representation. However, this representation is not convenient

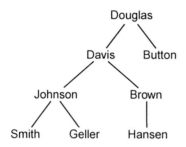

EmployeeID	Employee	ManagerID
1	Douglas	NULL
2	Davis	1
3	Johnson	2
4	Smith	3
5	Geller	3
6	Brown	2
7	Hansen	6
8	Button	1

Employee dimension table

Figure 3.5: A parent-child dimension and a relational representation of it

for analysis using standard SQL. If we want to find the total sales generated by the shop managers and every staff member below them, the representation in Figure 3.5 is very hard to use since there is no easy way to traverse the hierarchy and roll-up from assistants to shop managers. To overcome this, another representation with a so-called *bridge table* can be used. A bridge table holds a row for each path between an ancestor and a descendant in the hierarchy, including the trivial paths of length 0. The bridge table has the following columns.

- *Ancestor:* A foreign key column referencing the primary key column of the dimension table to capture a row's ancestor.

- *Descendant:* A foreign key column referencing the primary key column of the dimension table to capture a row's descendant.

- *Distance:* An integer capturing the length of the path between *Ancestor* and *Descendant*.

- *Bottom Flag:* A Boolean value indicating whether *Descendant* is at the lowest level (i.e., does not have descendants).

- *Top Flag:* A Boolean value indicating whether *Ancestor* is at the top-most level (i.e., does not have ancestors).

The parent-child hierarchy shown in Figure 3.5 is represented by the bridge table shown in Figure 3.6 (assuming the same EmployeeIDs as before). It can, e.g., be seen from the first row that Douglas with EmployeeID 1 manages himself; from the second row, it can be seen that Douglas also manages Davis who has EmployeeID 2.

The fact table remains unchanged, and it is possible to ignore the bridge table and join the fact table directly with the dimension table. In the example, this means that the fact table has a foreign key column that references the Employee dimension table and that the fact table represents what each individual employee has sold. The bridge table comes into play if we want to traverse the parent-child hierarchy and, e.g., roll-up sales to shop managers. To do that, we join the dimension, bridge, and fact table with the condition that EmployeeID (the primary key) of the dimension table

Ancestor	Descendant	Distance	Bottom Flag	Top Flag
1	1	0	False	True
1	2	1	False	False
1	8	1	True	False
1	3	2	False	False
1	6	2	False	False
1	4	3	True	False
1	5	3	True	False
1	7	3	True	False
2	2	0	False	False
2	3	1	False	False
2	6	1	False	False
2	4	2	True	False
2	5	2	True	False
2	7	2	True	False
8	8	0	True	False
3	3	0	False	False
3	4	1	True	False
3	5	1	True	False
6	6	0	False	False
6	7	1	True	False
4	4	0	True	False
5	5	0	True	False
7	7	0	True	False

Figure 3.6: A bridge table representing a parent-child hierarchy

is equal to the Ancestor attribute in the bridge table and the foreign key column in the fact table is equal to the Descendant column in the bridge table, and we place relevant constraints on the Employee dimension as shown next.

```
SELECT E.EmployeeID, SUM(F.Sales_Amount)
FROM Employee E, Bridge B, Sales F
WHERE E.EmployeeID = B.Ancestor AND B.Descendant = F.EmployeeID
AND E.Title = 'Shop Manager'
GROUP BY E.EmployeeID
```

To understand how this works, recall that the bridge table holds a row for each path between an ancestor and a descendant in the hierarchy. When we join the Employee dimension table (restricted to shop managers) to the bridge table, we get a row for each combination of a shop manager and an employee below the manager (including the manager). These rows are then joined with the fact rows such that we get the sales that the employees made. Finally, the GROUP BY ensures that we sum up all sales for employees below a given manager.

By using the Distance attribute in the bridge table, we can carry out more advanced analyses. If we add the constraint `B.Distance = 1` to the query above, we find the sales by those employees directly under the managers. If we instead use `B.Distance > 0`, we find the sales made by the managers' employees, but we do not include the sales the shop managers made themselves. Likewise, the Top and Bottom flags can be used to express some queries conveniently.

3.3.2   UNBALANCED HIERARCHIES

In an instance of an *unbalanced hierarchy*, dimension values exist that belong to a level different from the lowest level in the hierarchy's schema and do not have any children in the instance. In other words, dimension values for the lowest level are missing such that the instance forms a tree where the leaves have different distances to the ⊤ level.

One example of an unbalanced hierarchy is the parent-child hierarchy in Figure 3.5. Another example (which is not a parent-child hierarchy) arises if the book retailer's large shops become subdivided into *departments* such that departments belong to a shop. The smaller shops are not subdivided, meaning that a given shop has zero or more departments. The schema for the revised Shop dimension is shown in Figure 3.7, where we have extended the notation used so far to be able to show more advanced schemas.

Thus, a level is represented by a box where the level name is shown in bold on the top-most line. Below the level name and inside the box, level properties are shown, if any. A line between two levels A and B (where A is drawn above B) still means that there is a hierarchy where the A level is above the B level, i.e., B values are grouped into A values. Such a line can have explicit *cardinalities* shown where it connects to the boxes. In Figure 3.7, it can be seen that a shop has from 0 to many (the * represents an unbounded positive integer) departments, while a department belongs to from 1 to 1 (i.e., exactly 1) shop. If no cardinalities are shown, we implicitly assume that if A can roll-up to B then each A value belongs to exactly one B value and that at least one and possibly many A values belong to a given B value.

In Figure 3.8, which shows an instance of the revised Shop dimension, we can see that the shop Shop2 has departments while the shop Shop1 does not. (Ignore for now the part of the schema from City and up; we explain this part soon.)

At the implementation level, it is difficult to support unbalanced hierarchies. If the levels are of the same type, a parent-child hierarchy may of course be used. If the levels are of different types, it is, however, harder because different facts in effect have different granularities with respect to the hierarchy. One solution is to make the unbalanced hierarchy balanced by padding it with placeholder values at the lower levels. This operation can be done transparently by the system [67]. A complete coverage of this is beyond the scope of this book.

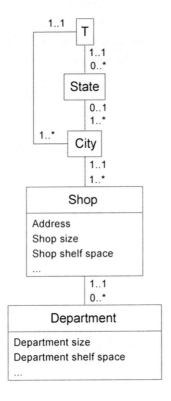

Figure 3.7: Revised schema for the Shop dimension

3.3.3 NON-COVERING HIERARCHIES

A *non-covering hierarchy* allows instances to skip levels between the leaves and the root. Consider the Shop dimension in Figure 3.7 where City values are grouped into State values. Here, the city Washington, D.C. does not belong to a state. This means that we skip the State level when representing Washington, D.C. This phenomenon would also occur if the book retailer expands to other countries without states, such as many European countries. Note that in our new notation, the cardinality 0..1 from City to State means that a City value may belong to no State value or to a single State value. If it belongs to no State value, it still belongs to a value from a level above State, in this case ⊤.

A way to support non-covering hierarchies is to insert placeholder values whenever a value lacks a containing value in the (immediate) parent level in the schema. In the example instance, a placeholder value is thus inserted at the State level. This placeholder value has the City level value Washington, D.C. as its single child. The name and level property values appropriate for this placeholder are determined by business needs. It may make sense to include a fake state called "Washington, D.C" such that analysts could do queries as if Washington, D.C. was a state, or it may be better to leave the name empty to show that this is just a placeholder. In some systems, such

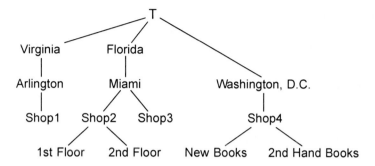

Figure 3.8: An instance of the revised Shop dimension

as Microsoft Analysis Services (where a non-covering hierarchy is called a *ragged hierarchy*), it is possible to hide placeholder values when an analyst makes use of a hierarchy.

3.3.4   NON-STRICT HIERARCHIES

We have so far considered hierarchies where each dimension value (apart from the special T value) belongs to exactly one parent value. Such a hierarchy is called a *strict hierarchy*. In an instance of a *non-strict hierarchy*, a child can have more than one parent such that many-to-many relationships can exist between dimension values at different levels. Non-strict hierarchies occur often in the real world. In our book retailer example, there is a many-to-many relationship between books and authors (an author can write many books, and a book can have many authors). This is shown in Figure 3.9.

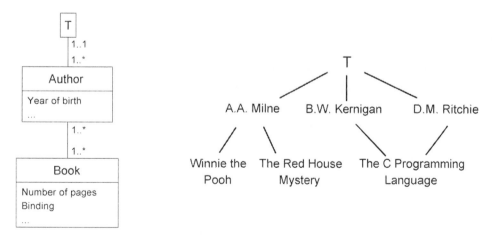

Figure 3.9: Schema and example instance for the Book dimension with many-to-many relations between authors and books

Many-to-many relationships cause problems with respect to aggregations. Assume that a book *B* is written by two authors. If we roll-up from the Book level to the Author level, and then further from the Author level to the ⊤ level, we count the sales of *B* twice (once for each author) unless we take special care.

To represent the many-to-many relationships in ROLAP, it is not enough to use a single column to represent a level in case of a star schema or a single foreign key in case of a snowflake schema. Instead, we include a table for each of the two levels participating in a many-to-many relationship and include a *bridge table* in-between the two. This is illustrated for the Book dimension in Figure 3.10.

BookID	Title	...
1	Winnie the Pooh	...
2	The C Programming Language	...
3	The Red House Mystery	...

Book level table

AuthorID	Name	...
1	A.A. Milne	...
2	B.W. Kernigan	...
3	D.M. Ritchie	...

Author level table

BookID	AuthorID
1	1
2	2
2	3
3	1

Bridge table

Figure 3.10: Tables representing the Book dimension in ROLAP

With the suggested solution, we cannot represent the ordering of authors of a book, and we cannot assume that authors are always ordered alphabetically. Therefore, we allow a many-to-many relationship to have *properties*, as illustrated in Figure 3.11. When represented in ROLAP, the properties of many-to-many relationships result in columns in the bridge table. In this example, the order is just a positive integer. Further, we include a percentage that shows how large a fraction is "owned by" a parent.

If we, e.g., roll-up the dollar sales to the Author level, this fraction is multiplied with the measure values such that we count an amount from a single sale exactly once. In Figure 3.12, we show an example of a book with two authors where the percentage is 0.5 for each of them. If a single copy of the book is sold for 40 dollars, and we roll-up the dollar sales to the Author level, we thus find that each of the two authors has generated a sales amount of $0.5 \cdot 40$ dollars = 20 dollars in that single sale. When loading data into the database, it is, of course, important that these property values are assigned correctly such that percentages add up to exactly 1.0 or that there, for a given book, is only one first author, etc.

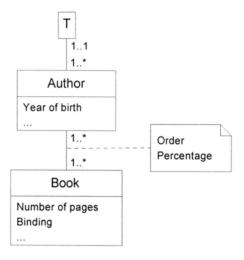

Figure 3.11: Schema for the Book dimension with many-to-many relations between authors and books and order represented

BookID	Title	...
1	Winnie the Pooh	...
2	The C Programming Language	...
3	The Red House Mystery	...

Book level table

AuthorID	Name	...
1	A.A. Milne	...
2	B.W. Kernigan	...
3	D.M. Ritchie	...

Author level table

BookID	AuthorID	Order	Percentage
1	1	1	1.0
2	2	1	0.5
2	3	2	0.5
3	1	1	1.0

Bridge table

Figure 3.12: Tables representing the Book dimension including properties in the bridge table

3.3.5 MULTIPLE HIERARCHIES AND PARALLEL HIERARCHIES

In many cases, it is convenient or even necessary to have more than one hierarchy in a dimension. Consider the revised Time dimension shown in Figure 3.13 where there are two hierarchies (level properties are not shown). Both group days into months, but they group months into quarters differently: One of the hierarchies represents the normal calendar, while the other represents a fiscal calendar.

Compared to Chapter 2 (cf. Section 2.2), we extend our notion of a dimension such that a dimension is organized into *one or more* containment-like hierarchies that may or may not share

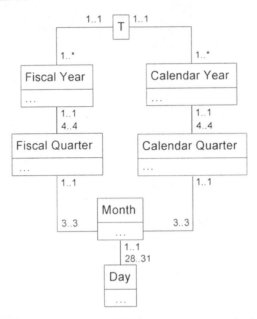

Figure 3.13: Schema for the Time dimension with multiple hierarchies

intermediate levels (the ⊤ level is always shared). When a dimension has more than one hierarchy, we say that the dimension has *multiple hierarchies*. Often, it does not make sense to use the multiple hierarchies together in a single analysis. In an analysis of the book retailer's sales, it is not interesting to consider a pivot table where the *x* axis shows fiscal quarters and the *y* axis shows calendar quarters. This is not interesting as the two hierarchies have the same analytical purpose: grouping of days into coarser time units. The hierarchies just group the days differently. Further, the groupings are very correlated, and a pivot table with these two groupings on the axes would only have cells with numbers along the diagonal.

In those cases where a dimension has two or more hierarchies that have different analytical purposes, we refer to them as *parallel hierarchies*. In the Book dimension for our book retailer, we have so far assumed that books are grouped based only on their author(s). However, it is also interesting for the book retailer to group books based on their genre and publisher. (Figure 3.14, to be covered shortly, illustrates this.) Hierarchies introducing these groupings have different analytical purposes, and it makes sense to use them simultaneously. For example, a pivot table can meaningfully show publishers on the *x* axis and genres on the *y* axis when the analyst investigates how well different publishers are doing with respect to different genres.

3.3.6   SUMMARIZABILITY

Consider a measure that is additive along a dimension. When higher-level aggregates can be computed from lower-level aggregates in a hierarchy in the dimension, we say that the hierarchy is *summarizable* with respect to the measure. For example, we can calculate our book retailer's nationwide sales of books from the sales of books in the different states, and the sales in states can be calculated from the sales in cities that can in turn be calculated from the sales in the individual shops. So, the hierarchy in the Shop dimension is summarizable with respect to the sales measure. In general, hierarchies should be made balanced, covering, and strict to ensure summarizability.

The ability to reuse already aggregated data in new queries has the potential for offering dramatic computational savings. It is much faster to aggregate (already calculated) numbers showing the sales in each state than it is to consider each fact when processing a query for the total number of sold books. Modern OLAP systems exploit such shortcuts heavily to provide fast answers based on precomputed aggregates.

3.4   SUMMARIZING EXAMPLE

We end the chapter by elaborating on the book retailer's data warehouse. We still consider the sales of books from shops as the business process. However, we use a transaction-level granularity and capture each individual sale. This results in a data warehouse with many more facts than with the design in Section 2.9 where sales where aggregated to show *daily* sales per book per shop. We use the same dimensions as in Section 2.9, but we extend them and also add some new dimension: Time of Day, Promotion, and Transaction.

Figure 3.14 shows the Book dimension, which has three parallel hierarchies such that the sales can be rolled up to authors, genres, and publishers. Note that a book can have several authors and several genres (such that bridge tables are needed in a ROLAP-based solution) and that these many-to-many relationships have properties. We decide that if a book has N authors (or genres) then the fraction for each author (or genre) should be $1/N$.

The Book dimension is not a slowly changing dimension; when a new book or book edition is published, a new dimension value is created.

The Shop dimension is shown in Figure 3.15. It has only one hierarchy. As we have discussed previously, the city Washington, D.C. does not belong to a state. If the book retailer has a shop there, we overcome this problem by pretending that Washington, D.C. is also a state and create a placeholder value. The Shop dimension is slowly changing, and type 2 updates are used. If we were to subdivide shops into departments to capture department-level sales, we would need to subdivide all shops into departments such that facts refer to departments. That means that for small shops, we would have to include a single, artificial department.

The Date dimension is shown in Figure 3.16. It has two (non-parallel) hierarchies that group days into normal calendar units and fiscal calendar units. Note that the Day level has a level property that is used to capture whether some special event takes place on a given day. This property should, however, be used only for events, e.g., Labor Day, that are relevant for all shops. If some local event

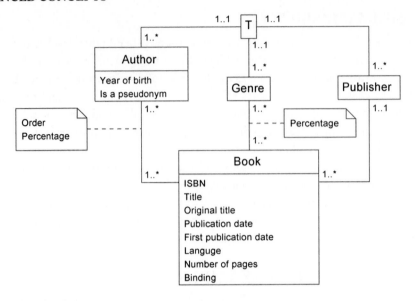

Figure 3.14: Book dimension

takes place near one shop, this event should not be represented in the Date dimension, as this would be misleading when analyzing sales from the other shops.

The Time of Day dimension is shown in Figure 3.17. It has a single hierarchy that groups minutes into hours and hours into periods ("Morning," "Noon," "Afternoon," "Evening," and "Night"). Note that the Time of Day dimension only makes sense with the finer granularity where we capture each individual sale. When we considered daily sales, we could not see when the sales happened during the day.

The Promotion dimension is shown in Figure 3.18. It represents how the book is displayed (e.g., "Displayed in Window"), whether it is discounted (e.g., "Clearance sales"), and whether some promotion takes place (e.g., "Book signing"). These differ from shop to shop and from book to book, so we cannot represent them in the Book or Shop dimensions. They are unrelated, so they could be placed in different dimensions, but as they all have few possible values, we group them together in a junk dimension.

The Transaction dimension is included to be able to see which books were sold together in a single sale. However, we have already captured where the transaction took place (with the Shop dimension) and when (with the Time of Day and Date dimensions). In other words, since there is nothing left to be represented in the Transaction dimension, we choose to model it as a degenerate dimension consisting only of an integer that uniquely identifies each transaction.

As measures, we pick sales price and sales count, both with SUM as the associated formula (although we consider the individual transactions, a customer can buy two or more copies of the same book).

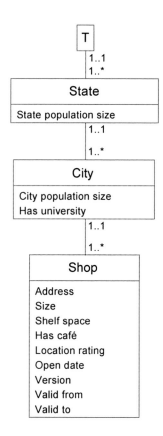

Figure 3.15: Shop dimension

3.5   EXERCISES

1. Extend the example in Section 3.4 to include information about the customers who are members of a loyalty program.

2. Consider the following Customer dimension.

ID	Login	Gender	Birthday	City	Credit rating	Version
1	Cat	Male	1980-11-02	Amsterdam	3	1
2	Fish	Male	1970-01-04	Berlin	4	1
3	Bird	Female	1980-05-02	London	5	1
4	Cat	Male	1980-11-02	Paris	3	2
5	Tiger	Female	1975-08-09	Madrid	4	1

Assume that each customer has a unique Login and that the dimension is slowly changing. Changes to Gender and Birthday are handled with type 1 updates (overwrites) while changes

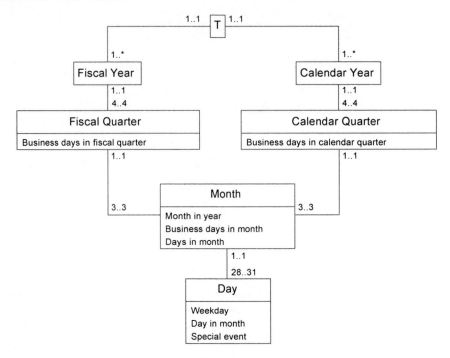

Figure 3.16: Date dimension

to City and Credit rating are handled with type 2 updates (new versions). Show the dimension after the following changes.

(a) The customer Cat corrects his birthday information to 1981-11-02.

(b) The customer Tiger moves to Amsterdam.

(c) The credit rating for the customer Tiger is updated to 5.

(d) The customer Tiger updates her birthday information to 1975-09-09.

3. Discuss if updates of the Login attribute can be handled in Exercise 2.

4. If type 2 updates are used for handling a slowly changing dimension, it is often the case that when a row version v_1 is replaced by a new row version v_2, all facts will refer to v_2. This is, for example, the case when a supermarket is expanded. After the expansion, all sales facts will refer to the Shop dimension's new row version showing the the expanded size. In other cases, new facts may, however, meaningfully refer to previous versions. Give an example of this.

5. Consider a chain of stores where the sales districts are reorganized each year (always with effect from January 1st). The management wants to be able to see which sales district a shop

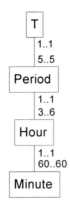

Figure 3.17: Time of Day dimension

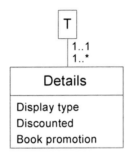

Figure 3.18: Promotion dimension

belonged to in a given year by means of the Shop dimension. Discuss if this can be supported by using the following.

(a) type 1 updates

(b) type 2 updates

(c) type 3 updates

(d) generalized type 3 updates

6. Recall your solution to Exercise 7 in Section 2.10. Now assume that the chain invests in new point of sale (POS) technology. The new POS system collects information about each individual sales transaction (a given transaction may involve more than one publication—e.g., the customer could buy both "Local News" and "Global News" at the same time) and include timestamp and salesperson. This information can be transferred automatically to the headquarters. Extend your solution to handle the finer data granularity and extended context. Include a degenerate dimension, an outrigger, and a junk dimension in your design.

7. Write SQL that uses the bridge table in Figure 3.6 to find the sum of the sales amounts from all sales made by Brown (with EmployeeID 6) or anyone *above* Brown in the organization.

8. Consider a parent-child hierarchy that captures the following: The company A1 owns the companies B1, B2, and B3. The company A2 owns the company B4. The company B1 owns the company C1.

   (a) Show the parent-child hierarchy instance graphically.

   (b) Show the parent-child hierarchy instance in the form of a bridge table.

9. How can the number of rows in a bridge table for a parent-child hierarchy be calculated?

10. Give examples of non-summarizability in

   (a) unbalanced hierarchies

   (b) non-covering hierarchies

   (c) non-strict hierarchies

11. For each of the following hierarchy instances, determine whether it is each of balanced, covering, and strict.

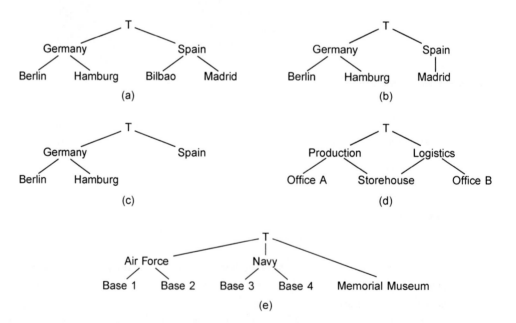

CHAPTER 4

Implementation Issues

In this chapter, we describe how multidimensional databases are implemented. Because multidimensional databases often hold hundreds of gigabytes or terabytes of data, smart solutions are needed to achieve good performance. When analysts use a multidimensional database, they should in most cases only need to wait seconds (or less) to get answers to their queries. The queries may, however, need to access and aggregate large amounts of data, e.g., to find the total sales in the last five years grouped by product category, and it is challenging to answer such queries quickly. Much of this chapter is therefore focused on query performance, but we also describe different ways to implement OLAP systems and the workings of the Extract-Transform-Load (ETL) process.

4.1   MATERIALIZED VIEWS

Materialized views that hold aggregates are one of the most important means of improving query performance in a multidimensional database management system. In a nutshell, a precomputed aggregate view is the materialized result of a query that aggregates (e.g., summarizes) parts of or all the data in the database. The word "materialized" means that the result is stored physically such that the system can answer the query by just reading the (already computed and stored) result instead of computing it from the detailed base data.

The use of materialized views can speed up query processing drastically to be several hundreds or thousands times faster than when using the base data. Often many different queries can even be answered efficiently by using the same materialized view. If we, for example, have an aggregate that holds the summarized sales for cities, we can also use that aggregate to calculate the total sales for states and the single total of all sales. It is important to note, however, that the end user should not be burdened with the task of deciding when to use a precomputed aggregate: The system has to take care of this transparently to the user.

In the following, we consider a ROLAP-based system with a star schema like the one shown in Figure 4.1 where SalesID is a degenerate dimension. The techniques we describe also work for more complex schemas like the one presented in Section 3.4, but for clarity, we use a simple example here.

With such a schema, most queries sent to the DBMS would be of the form shown next.

```
SELECT levels, SUM(Sale)
FROM the fact table  joined with one or more dimension tables
WHERE some conditions hold
GROUP BY levels
```

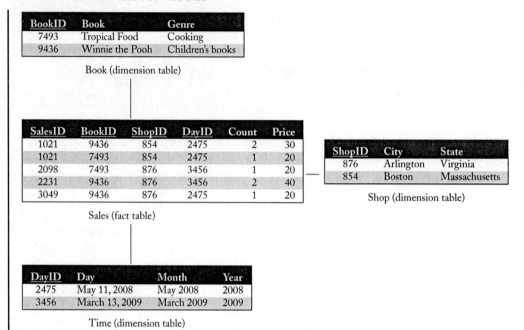

Figure 4.1: A star schema

To answer a query of this form can be very time consuming if all rows from the fact table need to be used to find the answer. Assume that the Sales fact table in Figure 4.1 has a billion rows, the Book dimension table has 100,000 rows, and the Shop dimension table has 100 rows. If our book retailer now wants to determine how the sales are distributed across genres, the following query can be used:

```
SELECT Genre, SUM(Price)
FROM Sales, Book
WHERE Sales.BookID = Book.BookID
GROUP BY Genre
```

To answer this query, we use all rows (one billion!) from the fact table.

Assume now that there is a precomputed aggregate that holds the total sales of each book from each shop. This materialized view can be created as shown next.

```
CREATE MATERIALIZED VIEW TotalSales(BookID, ShopID, Price, Count) AS
SELECT BookID, ShopID, SUM(Price), SUM(Count)
FROM Sales
GROUP BY BookID, ShopID
```

The considered query can then be rewritten (by the system—not by the user) to use the precomputed aggregate:

```
SELECT Genre, SUM(Price)
FROM TotalSales, Book
WHERE TotalSales.BookID = Book.BookID
GROUP BY Genre
```

At most (i.e., if each book was sold from each shop), the aggregate holds 10 million rows since we assumed there were 100 shops and 100,000 books. To read 10 million rows (which have even fewer columns) is much faster than reading a billion rows. As a rough estimate of the time it takes to execute a query, we can use the number of rows that must be read.

The problem is then to decide which precomputed aggregates to create. One extreme is to not have any. But if there are no precomputed aggregates, query responses will be slow. The other extreme is to precompute each possible aggregate. Unfortunately, they will occupy huge amounts of disk space and will take unreasonably long to compute in the first place. Further, they need to be updated when updates to the base data occur. So this extreme is also not practical. A good compromise in-between the two extremes must therefore be found. Unfortunately, the problem of picking the optimal set of aggregates is an NP hard problem.

In practice, heuristic techniques are used in picking a good (but likely not optimal) set of materialized views. As already stated, an aggregate can often be used to answer many queries. This then influences the decision of which aggregates to precompute. A good heuristic is to precompute an aggregate if it is considerably smaller than another precomputed aggregate that can be used. In one paper [31], a greedy heuristic algorithm is proposed. This algorithm is guaranteed to pick a set of aggregates with a *benefit* that is at least 63% of the benefit of an optimal algorithm (the benefit can be seen as the number of rows that do not have to be read when answering each possible query).

Modern OLAP systems support precomputed aggregates and implement good heuristics for choosing them. When choosing them, the tools can also take usage patterns into consideration such that the most often used queries get optimized the most. The use of the precomputed aggregates happens transparently to the end user when using such tools.

4.2   INDEXING

Indexing is another important means of improving query performance, and it is applied in ROLAP as well as in MOLAP. Here, we explain indexing from a ROLAP viewpoint.

4.2.1   INDEXING OVERVIEW

Fact tables are typically very large and are used in the majority of queries. Therefore, a fact table often has many indices such that most queries can benefit from one or more of them. Typically, a separate index is built on each dimension key. If the DBMS supports index intersection, the indices can then be used in combination when answering a query. It can also be a good idea, however, to create indices on combinations of dimension keys for those combinations that are often used together by queries. There are too many combinations to create an index on each combination (the order of

RowID	BookID	Title	Binding	Language	...
1	9436	Winnie the Pooh	Hardcover	English	...
2	1029	Le Petit Prince	Paperback	French	...
3	8733	Alice in Wonderland	Paperback	English	...
4	2059	Wind in the Willows	Hardcover	English	...
5	5995	A Bear Called Paddington	Paperback	English	...
6	1031	Pierre Lapin	Hardcover	French	...
7	3984	Le avventure di Pinocchio	Hardcover	Italian	...

Figure 4.2: Book dimension table also showing the special system-maintained attribute RowID, which normally cannot be seen by the user

the attributes in the index also matters). The combinations for which to create indices thus have to be chosen carefully based on the typical usage. In general, the key referencing the Date dimension should be put first in combinations since most queries refer to dates.

Dimension tables should also be indexed. Depending on typical usage, it may or may not make sense to build indices on all individual columns. And as for fact tables, indices can be built on combinations of attributes.

Virtually any DBMS supports the B-tree index which is a tree structure where the leaves contain lists of row IDs (where a row ID can be a physical location of the row or something else the system can use to identify and find a certain row). To find rows that have a given value for the indexed attribute, the system traverses the tree and obtains a list of row IDs. The row IDs are then used to retrieve the rows. The B-tree is an efficient index. However, when the indexed attribute has low cardinality (i.e., holds few different values), another representation can be a better choice. We describe this index type below.

4.2.2 BITMAP INDICES

When there are few values in a column, we can create a position bitmap for each value instead of maintaining lists of row IDs for each value (as in the B-tree). We call such a collection of position bitmaps a *bitmap index*.

Consider, for example, columns such as Binding and Language in the Book dimension table shown in Figure 4.2 where there are few distinct values. For ease of presentation, Figure 4.2 also shows the row IDs that correspond to the physical order of the rows in the table. A bitmap index on Binding has the following two position bitmaps:

```
Hardcover:  1001011
Paperback:  0110100
```

A bitmap index for Language has the following three position bitmaps:

```
English:  1011100
French:   0100010
Italian:  0000001
```

The principle of the bitmap index is that the position bitmap for the value v has a 1 on position p if and only if the p'th row in the table takes the value v. For example, the position bitmap for Hardcover in the bitmap index on Binding has a 1 in the first position, meaning that the first row in the table has the value "Hardcover" in Binding.

A bitmap index makes it very fast to locate rows with a certain value. If we were to find all books in English, the position bitmap for English would immediately indicate which rows to retrieve. It is also possible to combine bitmap indices. To find all hardcover books in English can be done by computing the logical AND of the position bitmaps for hardcover in Binding and for English in Language. In other words, we compute:

```
      1001011
AND   1011100
=     1001000
```

This tells us that the first and fourth rows in the table fulfill the query. The logical AND can be computed very efficiently as a single CPU operation can compare 32 or 64 bits on a modern CPU.

It is also possible to use other operators on bitmap indices. To find the books written in French or Italian, we compute the logical OR of the position bitmaps for French and Italian in Language:

```
      0100010
OR    0000001
=     0100011
```

It is also possible to find the books written in a language other than English (i.e., the books written in French or Italian in this example), by computing the logical NOT of the position bitmaps for English in Language:

```
NOT   1011100
=     0100011
```

When there are few values, a bitmap index occupies little space. There is one position bitmap for each distinct value, and in each position bitmap, a single bit is needed for each row. In contrast, the B-tree would need space to store a row ID for each row and some extra space for intermediate nodes. As a row ID typically consists of 4 or 8 bytes (i.e., 32 or 64 bits), it is clear that the bitmap index is advantageous when the cardinality of the indexed column is low.

It is also possible to make the bitmap index occupy less space. The first optimization is to skip trailing zeros (i.e., a sequence of zeros not followed by a one). If the table has x rows, but the length of the position bitmap for a given value is $y < x$, it can be assumed that the last $x - y$ positions in the bitmap hold zeros. Other optimizations can be achieved from various kinds of compression. In a bitmap index on an attribute with n different values, each position (each column in the two example bitmap indices shown above) in the position bitmaps has exactly one bit set to 1 and the remaining $n - 1$ bits set to 0. In other words, the probability of a 1 in an index is $1/n$. The larger n is, the smaller the probability of seeing a 1. In other words, there are many sequences consisting only of 0s. These sequences are very suitable for compression. A number of different algorithms exist for this kind of compression, and they help by making bitmap indices occupy less space while still being very efficient to use.

4.3   JOIN INDICES

A join index is another means of speeding up query performance. While it differs from materialized views and indices, it shares characteristics of both.

Like a materialized view, a join index represents the pre-computation of a query, in this case, a join query; and when updates occur to base data, a join index generally also needs to be updated to ensure that it remains correct. Unlike a materialized view, which is a self-contained unit of data, a join index contains pointers to the rows in the argument tables that satisfy the join predicate.

Like an index, a join index references base data using pointers and serves to reduce or eliminate searching for certain queries. As is typical for indices, a query that uses a join index will likely also need to access data outside the join index proper, namely the base data pointed to. Unlike an index and like a materialized view, a join index is capable of speeding up certain queries dramatically, while offering no speedup whatsoever of other queries. Specifically, queries that are covered by the join index can be supported, while the join index offers no support for non-covered or unrelated queries. An index such as a B-tree or a bitmap index in a sense supports a broader range of queries but with less dramatic speedups.

Consider, for example, the part of a star schema shown in Figure 4.3, which contains the fact table Sales and the dimension table Book. A common query pattern is to join Sales and Book to investigate how books sell. Without a join index, this would lead to a nested-loop join or a hash join. With the join index shown in Figure 4.3, however, the join is already computed and stored. For each row in Book, the join index holds a list of pointers to rows in Sales that join to that row in Book. For example, it can be seen that the row with RowID 1 in Book joins with the rows with RowIDs 2, 4, 5, and 8 in Sales. Note that the RowIDs are pointers to the rows in the tables allowing the rows to be retrieved quickly. Instead of storing lists of pointers to rows, position bitmaps can be stored.

By doing this, the join index looks as shown in Figure 4.4. It can be seen that the row with RowID 1 in Book joins with the rows at the 2nd, 4th, 5th, and 8th positions in Sales. These position bitmaps hold many more zeros than ones and should be stored in a compressed form. A join index can also be extended to include attributes, e.g., the join attribute BookID or attributes used for selection such as Genre.

A join index represents a different point in the range of tradeoffs between the use of storage space and query performance than do materialized views. A variety of algorithms may be used for updating join indices when updates to the base data occur. Join indices may be applied to one, two, or several tables.

RowID	BookID	Title	Genre	...
1	7493	Tropical Food	Cooking	...
2	9436	Winnie the Pooh	Childrens' books	...
3	9948	Gone With the Wind	Fiction	...
4	9967	Italian Food	Cooking	...

Book (dimension table)

RowID	BookID	ShopID	SalesID	DayID	Count	Price
1	9436	854	1021	2475	2	30
2	7493	854	1021	2475	1	20
3	9948	876	2098	3456	1	20
4	7493	876	2231	3456	2	40
5	7493	876	3049	2475	1	20
6	9436	854	3362	3569	2	30
7	9967	731	3460	3569	1	35
8	7493	731	3460	3569	1	15
9	9948	731	3460	3569	1	15

Sales (fact table)

Book_RowID	Sales_RowID
1	$\langle 2, 4, 5, 8 \rangle$
2	$\langle 1, 6 \rangle$
3	$\langle 3, 9 \rangle$
4	$\langle 7 \rangle$

Join index for Book and Sales

Figure 4.3: A fact table, a dimension table, and a join index with a list of pointers

4.4   QUERY PROCESSING

When processing queries, the system performs different optimizations to achieve fast processing. In Section 4.1, we saw how a precomputed aggregate can be significantly faster to use than the base data. It is the job of the DBMS, or a specialized middleware in case the DBMS cannot do it, to rewrite queries to use precomputed aggregates when possible. The user should not reference the precomputed aggregates in queries.

The DBMS will also decide which indices to use. RDBMSs generally support B-trees, and some also support bitmap and join indices as described in Section 4.2. As discussed there, the choice of indices also has a significant influence on the query execution time.

When choosing an execution plan for a query, an RDBMS can sometimes exploit the characteristics of a star schema. Recall that the dimension tables have very few rows compared to the fact table. A typical query joins the fact table with a number of dimension tables in a so-called *star join* and also places restrictions on the dimension tables. An example of such a query is given next.

RowID	BookID	Title	Genre	...
1	7493	Tropical Food	Cooking	...
2	9436	Winnie the Pooh	Childrens' books	...
3	9948	Gone With the Wind	Fiction	...
4	9967	Italian Food	Cooking	...

Book (dimension table)

RowID	BookID	ShopID	SalesID	DayID	Count	Price
1	9436	854	1021	2475	2	30
2	7493	854	1021	2475	1	20
3	9948	876	2098	3456	1	20
4	7493	876	2231	3456	2	40
5	7493	876	3049	2475	1	20
6	9436	854	3362	3569	2	30
7	9967	731	3460	3569	1	35
8	7493	731	3460	3569	1	15
9	9948	731	3460	3569	1	15

Sales (fact table)

Book_RowID	Sales_positions
1	010110010
2	100001000
3	001000001
4	000000100

Join index for Book and Sales

Figure 4.4: A fact table, a dimension table, and a join index with position bitmaps

```
SELECT Book, SUM(Count)
FROM Sales, Book, Date
WHERE Sales.BookID = Book.BookID AND Sales.DayID = Date.DayID
AND Book.Genre = 'Cooking' AND Date.Month = 'May 2010'
```

The query optimizer may recognize a star join and find plans that consider the (relatively small) dimension tables before the (big) fact table. More concretely, the DBMS can use the given predicates to first filter the dimension tables and then compute the Cartesian product of the remaining rows. This is in contrast to non-OLAP settings, in which the optimizer often tries to avoid computing Cartesian products.

In the example above, there are only 31 days in May 2010 and a limited number of cookbooks, say, 1,000. The Cartesian product with all combinations of days and books to consider thus only holds 31,000 rows. When this Cartesian product is computed such that the relevant key values for DayID and ShopID are available, the referencing facts can be retrieved efficiently from the fact table by means of indices.

Another method to compute a star join exploits join indices with position bitmaps like the one shown in Figure 4.4. This method first considers each dimension table to find rows that fulfill the given predicates. The join index's position bitmaps for all found rows are then combined by means of a logical OR. For the query above and the example in Figure 4.4, the Book dimension would thus be considered to find the two books with the genre Cooking. The join index's position bitmaps for these two books would then be combined as follows:

```
     010110010
OR   000000100
=    010110110
```

The resulting position bitmap shows which rows in the fact table refer to books with the genre Cooking. Something similar is done for the Date dimension to find a position bitmap for those fact table rows that refer to dates in May 2010. In the second step, the method then combines the position bitmaps found for each of the query's dimension tables by computing the logical AND of them. The resulting position bitmap shows which fact table rows that are relevant to the query.

In a third method to compute a star join, queries are rewritten to use subqueries such that the relevant fact table rows can be retrieved efficiently. The query above could, for example, be rewritten into the following query:

```
SELECT Book, SUM(Count)
FROM Sales, Book
WHERE Sales.BookID = Book.BookID AND Book.Genre = 'Cooking'
AND Sales.DayID IN
(SELECT DayID FROM Date WHERE Month = 'May 2010')
```

By first finding the relevant DayIDs from the Date dimension in a subquery, a bitmap index on the DayID attribute in the fact table can be used to efficiently reduce the number of fact table rows to consider. In the second step, the dimension tables with the attributes needed (only Book in this example) are joined with the result of the first step. A method like this is used by the Oracle DBMS [62].

4.5   OLAP IMPLEMENTATIONS

OLAP systems can be implemented in different ways. We have already seen how relational database technology can be used in so-called ROLAP systems. Here, we briefly present the dominant categorization of OLAP systems.

Multidimensional OLAP (MOLAP) systems store data in specialized multidimensional structures. A simple implementation is to use multidimensional arrays for storing a cube, but for a sparse cube (i.e., a cube with many empty cells), large parts of the allocated arrays would not hold data. MOLAP systems therefore typically include provisions for handling sparse arrays and apply advanced indexing and hashing to locate the data when performing queries.

In contrast, *Relational OLAP* (ROLAP) systems use relational database technology for storing the data. Section 2.5 describes how relational tables are used for dimension and fact data. In order to achieve good query performance, ROLAP systems employ specialized index structures such as bitmap indices, materialized views, and precomputed aggregates.

Generally, MOLAP systems provide faster query response and more space-efficient storage, while ROLAP systems scale better in the number of facts, are more flexible with respect to cube redefinitions, and provide better support for frequent updates. However, the boundaries are changing as MOLAP systems are becoming more scalable.

The virtues of the two approaches are combined in the *Hybrid OLAP* (HOLAP) approach, which generally stores higher-level summary data using MOLAP technology, while using ROLAP technology to store the detail data. Some HOLAP systems allow even more flexibility, giving users a choice between MOLAP and ROLAP at the level of individual cubes and/or materialized aggregates.

4.6 EXTRACT-TRANSFORM-LOAD

In this section, we describe the Extract-Transform-Load (ETL) process that is an important part of a data warehouse project. Recall that a data warehouse is a repository of integrated enterprise data collected from different source systems within, and possibly also outside, the enterprise. To get the data into the data warehouse, the ETL process is used. It consists of three main steps as indicated by its name.

In the *Extract* step, data from the source systems is extracted. The contents of a data warehouse typically stems from different heterogeneous sources like relational databases, flat files, spreadsheets, logs, and different applications used in the organization, e.g., ERP systems. As data warehouses hold large amounts of data, it is too time consuming to extract all the relevant data each time the ETL is run. Instead, the ETL should only extract data that was added or updated since last time the ETL was run. This "delta" is then used to update the data warehouse to represent the new state. In some cases, it is easy to find the delta. For example, a unique log file may exist for each day such that it is easy to locate the new data. In other cases, it is much more difficult. It could, e.g., be the case that the source system with customer information only holds the current data and cannot tell what was added or updated since the last extraction. In that case, the ETL process must compute the delta itself by comparing the available data with the previous extract.

In the *Transform* step, transformations are applied that make the data fit the schema of the data warehouse. This includes combining the data from the different source systems. Much of the work done in this step is *cleansing* where aspects of the data, e.g., dates, spellings, and codings, are homogenized and where data of a bad quality is fixed, e.g., by adding missing fields or correcting wrong values such as negative sales prices. Unfortunately, the quality of the data extracted from the source systems often does not live up to the strict requirements needed for the data warehouse, so much cleansing is needed. Another typical transformation is conversion of character sets, e.g., from EBCDIC to Unicode. The transformations must also normalize/denormalize the data into the desired dimensional data warehouse schema. It is also the job of this step to assign surrogate keys to the data and maintain slowly changing dimensions by detecting changes and taking the actions necessary to contend with these. It is important that ETL maintains referential integrity such that a fact does not refer to a non-existing dimension value. The ETL transformations must ensure that only data of good quality is loaded into the data warehouse. A popular saying about allowing bad data in the data warehouse is "Garbage in, garbage out."

In the *Load* step, the data warehouse is fed with new data such that the data structures (e.g., tables in case of relational technology) are updated to contain the new data. Normally, the data warehouse is taken offline during the load such that no users query the data warehouse concurrently. Since data warehousing involves large amounts of data, it can be challenging to load the data warehouse in a limited time frame. Again, it is much faster to handle a delta than doing a full load in each ETL run.

To speed up the loading, the data warehouse's indices can be dropped and rebuilt after the load (but before users start querying the data warehouse again). Parallel loads of different dimensions and fact tables represents another means to gain faster loading. In addition to loading the base data, it may also be necessary to update other structures, e.g., precomputed aggregates, join indices, or MOLAP cubes. The load traditionally takes place at a time when fewest users are bothered by it, e.g., during the weekend or during the late night. This means that the data warehouse does not hold the newest data but only data from the previous week or previous day. Recently, it has, however, become popular to do "near real-time data warehousing" where the data warehouse is refreshed frequently and—nearly—holds the newest data. This calls for new and advanced solutions, as the data warehouse obviously cannot be taken offline many times during a day to be loaded.

While the purpose and tasks of the ETL may sound to be well-understood and easy, the opposite is, in fact, the case. In virtually any data warehouse project, the majority of the time is spent on getting the ETL right. A lot of time is needed to understand when, where, and how to extract data from the different sources and to understand what the data from different sources mean and how the data can be combined. In addition to that comes all the technical challenges about obtaining fast loads, etc.

4.7 EXERCISES

1. Consider the star schema shown in Figure 4.1 and the materialized view TotalSales created as follows:

```
CREATE MATERIALIZED VIEW TotalSales(BookID, ShopID, Price, Count) AS
SELECT BookID, ShopID, SUM(Price), SUM(Count)
FROM Sales
GROUP BY BookID, ShopID
```

For each of the following queries, show how it can be rewritten to use TotalSales or explain why it cannot be rewritten to use TotalSales.

(a) SELECT State, SUM(Count)
    FROM Sales, Shop
    WHERE Sales.ShopID = Shop.ShopID
    GROUP BY State

(b) SELECT Book, MAX(Price)
    FROM Sales, Book
    WHERE Sales.BookID = Book.BookID AND Book.Genre = 'Cooking'
    GROUP BY Book

(c) SELECT Year, SUM(Price)
    FROM Sales, Time
    WHERE Sales.DayID = Time.DayID
    GROUP BY Year

(d) SELECT State, Genre, SUM(Price), SUM(Count)
    FROM Sales, Book, Shop
    WHERE Sales.ShopID = Shop.ShopID AND Sales.BookID = Book.BookID
    GROUP BY State, Genre

2. Consider the following Employee dimension:

EmployeeID	Name	Gender	Office	Title
1	Bruce	Male	London	Developer
2	Carl	Male	Glasgow	Tester
3	Dorthea	Female	Cardiff	Developer
4	Erica	Female	London	Tester
5	Frederic	Male	London	Project manager
6	Gina	Female	Glasgow	Developer
7	Harvey	Male	Cardiff	Developer

Show the bitmap indices on the Gender, Office, and Title attributes and use them to:

(a) Find all female developers.

(b) Find all employees located outside Cardiff.

(c) Find all female employees located outside Cardiff.

(d) Find all employees that are developers or working in the London office.

3. Bitmap indices can also be used on numeric attributes such as Salary. However, a position bitmap is typically not made for each possible value then. Instead, one uses the "binning technique" where values are grouped together in "bins": [0; 1000), [1000; 2000), [2000; 3000), etc.

(a) Explain why a position map is not made for each value taken by a numeric attribute.

(b) Show the bitmap index for Salary in the following table by using bins of size 1000 as listed above.

RowID	EmployeeID	...	Salary
1	40	...	980
2	45	...	1600
3	100	...	900
4	23	...	2000
5	55	...	1500
6	69	...	2500
7	20	...	1100
8	87	...	2999

(c) Explain how the bitmap index can be used to find those employees with a salary between 1500 and 2500. Do the position bitmaps give the answer directly or is a refinement step needed?

4. Show a join index for Sales and Shop in Figure 4.1.

5. Show the details of the different ways to compute the result of the following star join query on the data in Figure 4.1.

```
SELECT State, SUM(Price)
FROM Sales, Book, Shop, Time
WHERE Sales.BookID = Book.BookID AND Sales.ShopID = Shop.ShopID
AND Sales.DayID = Time.DayID
AND Time.Year = 2008 AND Book.Genre = 'Cooking'
GROUP BY State
```

CHAPTER 5

Further Readings

This chapter first offers an overview of background material for the coverage offered in the preceding chapters, including material on which those chapters are based. Then Section 5.2 briefly covers readings on a range of advanced topics that we have not covered in any detail, but that represent extensions of the subjects covered and that are subject to ongoing research. Finally, Section 5.3 offers an overview of software resources that are available for free download.

5.1 BACKGROUND READINGS

Through its coverage of general background material, this section offers context for those who want to study topics covered in the previous chapters in further detail, and it identifies the sources on which the previous chapters are based.

History of the Field   As noted in Section 1.2, industrial OLAP tools date back to as early as the late 1960s. Data warehouses started to appear in industry in the 1980s, at which time the first companies began to build large data warehouses that integrated data from many sources for analysis purposes.

Also in the 1980s, multidimensional databases became subject to academic research in the database community, where they were named *statistical databases* [80] because these databases were considered in the context of statistical applications. Data Warehousing emerged only as an academic subject in the mid 1990s [95]. The field then enjoyed an explosion of activity the next few years, with lots of work on data modeling, query language constructs, change propagation, materialized views, bitmap indices, etc.

The field is thus characterized by a dual nature, with significant concurrent contributions from academia and industry. In the following, we therefore reference both academic and practitioner-oriented work, as appropriate. A detailed account of the academic history of multidimensional databases and data warehouses, along with a discussion of many advanced topics such as hierarchies, time, materialized views, incomplete information, etc. is given by Rafanelli [75].

Inmon et al.   The person most often credited for coining the term "data warehouse" is Bill Inmon, who is also called "the father of data warehousing." The first edition of his bestselling book "Building the Data Warehouse" [35] was published in 1992 and has since appeared in new editions. In 2007, Computerworld listed Inmon among "The Ten IT People Who Mattered in the Last 40 Years" in the computer profession, citing him as the father of a 28 billion dollar industry.

Section 2.6, on data warehouse and data mart architecture, is inspired by Inmon's writings. Unlike Kimball, Inmon argues that a data warehouse should not be modeled exclusively using dimensional techniques. Rather, the data warehouse itself should be used only for data integration and consolidation, not for the actual data analysis, which should, instead, be done in separate (multi-dimensional) data marts based on the data warehouse [36]. Since multidimensional modeling can lead to an unfortunate and less general data warehouse design, Inmon argues that a data warehouse should be modeled using traditional relational database design techniques since the resulting designs are then more robust to future changes.

Inmon also advocates the so-called *operational data store* that, like the data warehouse, is an integrated and consolidated database containing enterprise data [37], but it is used only for operational purposes and thus contains only current data, not historical data. Most recently, Inmon and colleagues have published books about the next generation of data warehouses [39] and the use of unstructured data in data warehouses [38].

Kimball et al.   Ralph Kimball is a very influential pioneer in *dimensional modeling* of data warehouses based on relational databases, i.e., the relational OLAP approach. Kimball founded Red Brick Systems in 1986 and served as the company's CEO until it was sold to Informix, now owned by IBM, in 1992. He then entered the data warehousing consultancy and education business.

The first edition of Kimball's bestselling book on relational dimensional modeling "The Data Warehouse Toolkit" [48] was published in 1996 were it fueled the increasing industry interest in data warehouse modeling. The book contains a large number of case studies from different domains that serve to introduce new modeling concepts in an intuitive manner. According to Kimball, the dimensional modeling concepts of hierarchical dimensions and measures (which Kimball calls "facts") were first used by syndicated data providers such as ACNielsen that used them to integrate and analyse retail market data.

Kimball and co-authors have later published another best-selling book "The Data Warehouse Lifecycle Toolkit" [50] that covers the entire development and deployment lifecycle of a data warehouse, including initial requirements analysis, modeling, technical design, ETL, end-user applications, deployment, and continuous growth of the data warehouse system.

Another popular book [47] by Kimball and colleagues concerns specifically ETL and delves into advanced ETL techniques such as the near-real-time feeding of a data warehouse. Kimball and his so-called "Kimball Group" have published widely in industry magazines and blogs over the years on best practice and guidance for data warehouses. An updated collection of these is available [49].

The coverage of relational OLAP modeling in Sections 2.5 and 2.7, parts of Section 2.8, Sections 3.1–3.2, and parts of Section 3.3 is to varying degrees either inspired by or based on the writing of Kimball and his colleagues.

Multidimensional OLAP   Whereas Kimball has been a proponent for relational OLAP systems, Erik Thomsen has focused on "pure" multidimensional OLAP systems. His 1997 bestselling book "OLAP Solutions: Building Multidimensional Information Systems" describes multidimensional concepts and modeling in practical terms, detailed in a large number of case studies. Later, Thomsen and colleagues have published books that combine general concepts with tool-specific advice and information, e.g., "Microsoft OLAP Solutions" [89]. Another book with the same kind of focus is "MDX-Solutions: With Microsoft SQL Server Analysis Services 2005 and Hyperion Essbase" by George Spofford et al. [82] that provides an in-depth description of the MDX language covered briefly in Section 2.8.6.

Tool-Specific Books   A number of books are available that target users of specific business intelligence tools. Here we cover a few. Section 5.3, which covers software resources, provides information on how the tools covered here can be downloaded for use.

Sivakumar et al. [32] offer quite comprehensive coverage of Microsoft Analysis Services, covering aspects such as the design of dimensions and cubes, administration, and performance optimizations. The book's use of easy-to-follow, step-by-step explanations, complete with graphical illustrations, makes this a very helpful do-it-yourself book.

A book by Stackowiak et al. [83] concerns data warehousing using Oracle's offerings. The book's objectives are to offer best practices for designing, implementing, and managing data warehouses with Oracle technology, to enable integration with solutions from other vendors, to explain how to analyze data using Oracle's tools, and to help the reader address performance challenges. The book may serve well as an introduction to the offerings from Oracle.

Next, Ballard et al. [4] offer a brief overview of IBM's InfoSphere technology and delve into one component, the InfoSphere Warehouse Cubing Services, that is designed to provide a multidimensional view of data stored in a relational database. This book is available for free.

Finally, a book by Bouman and van Dongen [8] covers Pentaho's business intelligence offerings. It explains system installation, it covers key business intelligence and data warehouse concepts and techniques, and it explains the use of all parts of Pentaho's business intelligence suite.

Database Textbooks with Data Warehousing Coverage   The topics of multidimensional data modeling and data warehousing have also found their way into popular database textbooks. While the textbooks cover these topics in less detail than we do, we include here an overview of the textbook coverage.

The textbook [14] by Connolly and Begg stands out as one of the two textbooks with the best coverage of multidimensional databases and data warehousing. Specifically, one chapter offers an overview of data warehousing, a following chapter focuses on the design of a dimensional data warehouse based on star schemas, and a third chapter offers an overview of OLAP, i.e., the querying of a data warehouse. The textbook by Hoffer et al. [33] covers the motivation for data warehousing and the architecture of the data warehouse in quite some detail. Likewise, ETL is covered better than in other textbooks, as is star schema based data warehouse design.

Ramakrishnan and Gehrke [77] cover star schema design and offer the best coverage across database textbooks of querying using primarily SQL:1999. The textbook by Atzeni et al. [3] covers the architectural setting of a data warehouse, star and snowflake schemas, and querying. In their textbook [20], Elmasri and Navathe also include a chapter that offers an overview of data warehousing and OLAP, covering cubes and their relational embedding using star schemas. Next, Kifer et al.

include a chapter on OLAP and data mining in their book [46]. This chapter covers star schemas and the querying of these, and it covers implementation and ETL very briefly. Similarly, Date's textbook [15] affords data warehousing a brief coverage, thus covering ETL, architectural aspects, star schemas, and querying. Kroenke [51] covers data warehousing as a component of a larger enterprise system that aims to support reporting. He motivates data warehousing and explains its architectural context, and he touches upon OLAP as part of this context.

Chapter 4 covers a range of implementation-related techniques that are particularly relevant for multidimensional databases but that also have more general relevance. Thus, database textbooks [3; 14; 22; 46; 77; 81] now offer some coverage of techniques such as bitmap indexing, materialized views, and join indexes.

Industry Surveys The Business Application Research Center, BARC, an independent software industry analysis company, publishes the largest existing survey on business intelligence products, called the BI Survey, available at `www.bi-survey.com`, as well as a collection of in-depth business intelligence product evaluations, called the BI Verdict (previously called the OLAP Report), available at `www.bi-verdict.com`. These web sites offer some free information, while the full reports are available at a cost only. Readers with an industrial focus may find the free information on these websites informative.

Real Data Warehouses and Business Intelligence Systems We want to alert the reader's attention to Project REAL (where "REAL" is an acronym that expands into "Reference implementation, End-to-end, At scale, and Lots of users"). In this highly interesting and well documented project, Microsoft collaborated with several technology partners to build on actual customer scenarios to discover best practices for creating applications of multidimensional technology. The project, described at `http://www.microsoft.com/sqlserver/2005/en/us/project-real.aspx`, is based on Barnes & Noble's data warehouse and presents best practices for the development of a Business Intelligence system. Reports with a wealth of information, as well as a reference implementation, are available for download via this URL.

The world's arguably most famous data warehouse is that of Walmart, the largest grocery retailer in the US and the world's largest public corporation in terms of revenue (2010 figures). Paul Westerman, who was member of the team that built the Walmart data warehouse, offers insight into Walmart's data warehouse and the development of data warehouses, with focus in retail [94].

Scientific Literature Some of the content of the previous chapters are based in part on contributions from the scientific literature. Specifically, Pedersen et al. [65; 68] present a multidimensional data model that supports dimensions with multiple hierarchies that may be unbalanced, non-covering, and non-strict; our coverage of hierarchies in Section 3.3 is based in part on this work.

Section 5.2, next, covers readings drawn from the scientific literature more generally.

5.2   ADVANCED TOPICS

Due to the popularity of data warehousing, much research continues to be devoted to the extension of multidimensional data models with the objective of supporting a wider range of applications. This section covers selected advanced topics in this regard, thus offering pointers to further readings that may be helpful in identifying directions for further research.

Multidimensional Models for Complex Data   The sudden academic interest in multidimensional databases and data warehouses in the second half of the 1990s resulted in a numerous proposals for multidimensional data models. The data cube by Gray et al. [28] provided a formal basis for advanced aggregate querying of relational data, but it did not consider more advanced issues.

Other research focused on handling more complex multidimensional data such as complex and irregular dimension hierarchies (non-strict, unbalanced, attribute inheritance, etc.) and on complex fact-dimension relationships, and it considered the symmetric treatment of dimensions and measures [2; 11; 16; 29; 41; 52; 54; 65; 76; 92]. Real data warehouses may often benefit from the resulting support for complex data [97]. A related issue is the concept of *summarizability* [34; 53] that concerns whether or not lower-level aggregate values can be combined, e.g., summed up, to produce higher-level aggregate values. This concept relates to dimension hierarchies: aggregating over simple hierarchies (which form balanced trees) is summarizable if the associated aggregation function is distributive.

Several surveys [68; 93] offer detailed overviews of the multitude of multidimensional data models and their capabilities. Additionally, several recent textbooks [24; 56] discuss advanced modeling at the conceptual and logical levels, along with physical and implementation aspects of data warehouses.

Incomplete Multidimensional Data   For real-world data cubes, it is very common that the data is not perfect, but somehow "incomplete." Within data management in general, *incomplete data* is a term used to cover a number of data deficiencies and approaches to coping with these, including missing/unknown values, uncertainty, imprecision, disjunctive alternatives, exotic incomplete values (maybe values, open values, no information values, possible information), probabilistic values, and possibilistic (fuzzy) values [19]. A comprehensive survey of the treatment of incomplete information in multidimensional databases [19] groups the problems into four main categories: incomplete measures, incomplete grouping attributes, incomplete information in the hierarchy, and incomplete metadata.

A number of proposals for handling these issues exist. Dyreson [18] considers a data cube containing regions of unknown values. Queries on unknown regions are then either redirected to the nearest regions with complete information or computed as well as possible along with a completeness measure showing how much complete information is used in the query evaluation. Jagadish et al. [41] present techniques for handling problems in the metadata specifications and for aggregating when imprecise values in the grouping attributes exist. Pedersen et al. [66] propose techniques for handling both imprecise grouping values as well as imprecise measure values. Timko et al. [90] considers uncertain/imprecise data cubes for the special domain of spatio-temporal data, which introduce several new challenges. Finally, Barbara et al. [5] were the first to consider the related problem of compressing data cubes into approximate cubes, termed *quasicubes*, where incompleteness is introduced in return for reduction in storage space.

Temporal Data Warehousing   In Inmon's definition, a data warehouse is time variant and non-volatile. Thus, a data warehouse accumulates data and invariably has at least one temporal dimension. Facts reference values drawn from the time dimension(s) to capture some temporal aspect of the data. In the ideal case, the passing of time simply results in new facts being entered into the data warehouse, with the dimensions remaining static. However, things are more complicated in reality. We thus covered temporal aspects of a data warehouse, including the handling of so-called "slowly changing dimensions" that result when dimensions change over time (see Section 3.1), and we covered the design of time dimensions (see Section 3.2.5).

Owing to an active research community, the area of temporal relational databases was well established by the time dimensional data warehousing came into existence. Thus, quite a few temporal conceptual models and temporal data models and query languages [10] along with a foundation of temporal database concepts [43; 44] predate dimensional data warehousing.

In fact, temporal and multidimensional databases have evolved separately and with little cross fertilization. The former area was rooted primarily in academia, and its focus has roughly been on extending SQL with constructs that enable the effective querying of general temporal data. In contrast, the latter area was initially rooted primarily in industry, and it adopted a modeling approach to the handling of specifically temporal multidimensional data. One or more appropriately designed time dimension tables were used to enable complex aggregate queries that would otherwise be very difficult to formulate.

However, research has been conducted with the aim of enriching multidimensional data models and query languages with concepts from temporal databases.

In early research, Bliujūtė et al. [7] propose the concept of temporal star schemas that omit the time dimension table and, instead, timestamp each row in every table of the schema, treating the fact table and the dimension tables equally with respect to time. Pedersen et al. [65; 68] propose a multidimensional data model and query algebra supporting the standard temporal database concepts of valid and transaction time. In a series of papers, Vaisman and Mendelzon unfold a temporal multidimensional data model and query language [57; 58; 91]. Golfarelli and Rizzi [25] present a recent survey of research on temporal data warehousing that offers additional references.

Spatial and Spatio-Temporal Data Warehousing   Research has also been conducted on the extension of multidimensional data models and data warehousing to the geographical domain. Starting prior to year 2000, Bedard and colleagues have proposed, explored, and developed the concept of Spatial OLAP in a substantial and growing number of papers (e.g., [78]). Gómez et al. [27] present a recent survey of the Spatial OLAP literature and also delve into Spatio-Temporal OLAP where the objective is to support OLAP for moving objects [42]. The book on advanced data warehouse design by Malinowski and Zimányi [56] also covers spatial and temporal aspects.

Real-Time Data Warehouses   When integrating (mostly internal) data into the data warehouse, the typical setup has been to load data into the data warehouse periodically, e.g., every week or every night. However, in recent years, there has been an increasing trend towards requiring new data to be loaded into the data warehouse almost immediately after being created in the source systems [9].

This is typically termed *real-time data warehousing*, although a more accurate term is *near-real-time data warehousing*, since such systems do not have the ultrashort and ultrastrict timing requirements of a few milliseconds, which are often associated with "real" real-time systems. Instead, the requirement is that data is loaded as fast as possible, with delays of up to a few seconds or so.

It is often infeasible to keep a data warehouse (nearly) up-to-date with *all* available data, which has led to the notion of *right-time data warehousing* [87]. Here, the idea is to divide the data into several priority classes, so that the up-to-date'ness of a data warehouse is dependent on the priority of the data. This way, the data update frequency for a particular kind of data depends on the needs.

Within right-time data warehousing, research has been done on how to provide fine-grained data insertions with immediate data availability, while still maintaining the fast load speeds offered by bulk loading [87]. Research has also been done on how to schedule the data warehouse loading activities so that freshness and performance are balanced [85] and on how to include even more objectives into the scheduling of the real-time data warehouse loading [17; 84].

Loading of near-real-time data warehouses is often done using so-called *continuous data loading* where the transactions from the source systems are re-ordered and grouped before being propagated to the data warehouse [55]. Typically, near-real-time data warehouses are parts of larger real-time business intelligence systems [1; 17; 79] that typically combine stored warehouse data with aspects of (non-stored) data streams [12; 96].

Integration of External Data   Another interesting direction is the inclusion and integration of external data sources into multidimensional databases. We do not consider general-purpose data integration, which is an entire field in its own right [30], but instead focus on the work that specifically address multidimensional cubes.

The first work in this area considers how to virtually extend an existing (internal) multidimensional data cube with external dimensions from external object databases [64]. Today, most external data for integration into multidimensional data cubes is found on the web. A recent, comprehensive survey of integration of data warehouses with web data [73] groups the research in this area along three main lines:

i Using XML technology for integrating (regular multidimensional) distributed data warehouse systems.

ii Utilizing XML data in data warehouses in various ways.

iii Data warehouses queried in combination with document-centric XML collections.

The first line of research includes work on XML formats for exchanging multidimensional data and metadata, e.g., XML for Analysis [59] and on architectures for applying such formats and other XML technologies [61]. Work in the second line considers how to virtually extend an existing (internal) multidimensional data cube with external dimensions and/or measures from external XML data [63; 69]; other work studies the creation of creation of new cubes solely from XML data [26; 45; 74]. Finally, work in the third line aims to extend multidimensional cubes with a context of XML-based textual documents, termed *contextualizing* the cubes [71; 72].

A recent development is to integrate semantic web data into data cubes. Here, work has started on how to build virtual multidimensional data warehouses based on composing selected parts of ontologies located on the web [60].

5.3   SOFTWARE RESOURCES

In this section, we provide pointers to multidimensional systems that are available for free and can be downloaded from the Internet. These systems provide the reader to gain practical, hands-on experience with the concepts covered in this book and to learn how the concepts are implemented in real systems.

While the commercial tools dominate the business intelligence field, a large number of free open-source tools have appeared in recent years. One paper [86] presents a survey of open source ETL tools, DBMSs, OLAP servers, and OLAP clients, covering each individual tool, system, server or client. Another paper [23] compares "suites," i.e., bundles of integrated software that offer complete BI solutions. We now proceed to present commercial tools and after that open source tools. In doing so, we cover some of the most well-known tools and packages, but other less known tools do also exist.

Commercial Systems   The most widely used systems are commercial and thus cost money to use. However, some of these are available in free trial or development versions. Users should pay close attention to the different licenses to understand how the systems can be used.

IBM's InfoSphere Warehouse software can be downloaded from `www-01.ibm.com/software/data/infosphere/warehouse/enterprise.html`. The software includes the DB2 DBMS, an OLAP server, and query and reporting tools.

Microsoft's SQL Server can be downloaded from `www.microsoft.com/sqlserver/2008/en/us/default.aspx` (students may also be able to download it from `dreamspark.com`). SQL Server is bundled with an ETL tool, an OLAP server, and a reporting engine.

Oracle Database 11g is available from `www.oracle.com/technetwork/database/enterprise-edition/overview/index.html`. The DBMS provides built-in support for OLAP and cubes without the need for other servers.

TARGIT BI Suite can be downloaded from `www.targit.com/Products/Demo_and_Videos/Download_30_days_trial_demo.aspx`. It provides a graphical OLAP client for analysis and reporting. The tool emphasizes user-friendliness.

Open Source Extract-Transform-Load Tools   The ETL program Pentaho Data Integration (also referred to as PDI and Kettle) can be downloaded from `kettle.pentaho.com`. Another ETL tool called Talend Open Studio is available from `www.talend.com`. In both Pentaho Data Integration and Talend Open Studio, the user defines the ETL flow in graphical environments. Both Pentaho Data Integration and Talend Open Studio are available in "community editions," which are free to use and where the source code is available. They are, however, also available in commercial editions with more functionality.

Open Source DBMSs   A number of open-source relational DBMSs are also available. Two of the most interesting ones are, in our opinion, Ingres Database, available from `ingres.com`, and PostgreSQL, available from `postgresql.org`. Both have rich feature sets and are very solid.

Another open-source DBMS is LucidDB, available from `luciddb.org`. Custom-built for data warehousing and business intelligence, LucidDB supports advanced features such as star joins and bitmap indices. Further, it is a so-called *column-store* that, unlike traditional *row-stores*, stores the data in a table in a by-column fashion rather than in the traditional by-row fashion. This storage organization can be very efficient for data warehousing. However, at the time of writing, LucidDB still lacks other important features such as support for foreign keys.

Open-Source OLAP   Mondrian is the by far most popular open-source OLAP server. It can be downloaded from `mondrian.pentaho.com`. It supports MDX queries and is used in ROLAP environments such that a cube is defined on top of an underlying relational data warehouse. Mondrian can be used with any DBMS for which a JDBC driver exists. Next, JPivot, available from `jpivot.sourceforge.net`, is an OLAP client that presents data in pivot tables and graphs.

Open-Source Business Intelligence Suites   Three popular open-source business intelligence suites exist. JasperSoft Business Intelligence Suite, available from `www.jaspersoft.com/jaspersoft-business-intelligence-suite/`, comes with the MySQL DBMS and specialized versions of Talend Open Studio, Mondrian, and JPivot. Next, Pentaho Open BI Suite, which can be downloaded from `pentaho.com`, includes Pentaho Data Integration, Pentaho Analysis (using Mondrian and JPivot), as well as a dashboard application and a data mining application. Finally, SpagoBI, available from `www.spagoworld.org/xwiki/bin/view/SpagoBI/`, is an integration platform that provides the glue that allows the user to integrate different tools into a business intelligence solution. For example, Mondrian or Microsoft Analysis Services can be used as the OLAP server in a SpagoBI business intelligence solution.

Bibliography

[1] D. Agrawal. The Reality of Real-Time Business Intelligence. In *Proc. 2nd Int. Workshop on Business Intelligence for the Real-Time Enterprises*, pp. 75–88, 2008. 79

[2] R. Agrawal, A. Gupta, and S. Sarawagi. Modeling Multidimensional Databases. In *Proc. 13th Int. Conf. on Data Engineering*, pp. 232–243, 1997. DOI: 10.1109/ICDE.1997.581777 77

[3] P. Atzeni, S. Ceri, S. Paraboschi, and R. Torlone. *Database Systems—Concepts, Languages and Architectures*. McGraw Hill, 2000. 75, 76

[4] C. Ballard, D. Rangarao, J. Tang, P. Wittann, Z. Zakharian, A. Perkins, and R. Frankus. *InfoSphere Warehouse: Cubing Services and Client Access Interfaces*. IBM Redbooks, Vervante, 2008. Available from www.redbooks.ibm.com/abstracts/sg247582.html. 75

[5] D. Barbar and M. Sullivan. Quasi-Cubes: Exploiting Approximations in Multidimensional Databases. *ACM SIGMOD Rec.* 26(3):12–17, 1997. DOI: 10.1145/262762.262764 78

[6] K. S. Beyer and R. Ramakrishnan. Bottom-Up Computation of Sparse and Iceberg CUBEs. In *Proc. ACM SIGMOD Int. Conf. on Management of Data*, pp. 359–370, 1999. DOI: 10.1145/304181.304214 27

[7] R. Bliujūtė, S. Šaltenis, G. Slivinskas, and C. S. Jensen. Systematic Change Management in Dimensional Data Warehousing. In *Proc. 3rd Int. Baltic Workshop on DB and IS*, pp. 27–41, 1998. 78

[8] R. Bouman and J. van Dongen. *Pentaho Solutions: Business Intelligence and Data Warehousing with Pentaho and MySQL*. Wiley, 2009. 75

[9] R. M. Bruckner, B. List, and J. Schiefer. Striving towards Near Real-Time Data Integration for Data Warehouses. In *Proc. 4th Int. Conf. Data Warehousing and Knowledge Discovery*, pp. 317–326, 2002. DOI: 10.1007/3-540-46145-0_31 79

[10] M. H. Böhlen, J. Gamper, and C. S. Jensen. How Would You Like to Aggregate Your Temporal Data? In *Proc. 13th Int. Symp. Temporal Representation and Reasoning*, pp. 121–136, 2006. DOI: 10.1109/TIME.2006.17 78

[11] L. Cabibbo and R. Torlone. Querying Multidimensional Databases. In *Proc. 6th Int. Workshop on Database Programming Languages*, pp. 319–335, 1997. 77

[12] M. Castellanos, C. Gupta, S. Wang, and U. Dayal. Leveraging web streams for contractual situational awareness in operational BI. In *Proc. 2010 EDBT/ICDT Workshops*, 8 pages, 2010. DOI: 10.1145/1754239.1754248 79

[13] E. F. Codd. Providing OLAP (on-line analytical processing) to user-analysts: An IT mandate. E.F. Codd and Assoc., 1993. 2

[14] T. M. Connolly and C. E. Begg. *Database Systems—A Practical Approach to Design, Implementation, and Management*. 4th Edition. Addison Wesley, 2005. 75, 76

[15] C. J. Date. *An Introduction to Database Systems*. 8th Edition. Addison Wesley, 2004. 76

[16] A. Datta and H. Thomas. A Conceptual Model and Algebra for On-Line Analytical Processing in Decision Support Databases. In *Proc. 7th Annual Workshop on Information Technologies and Systems*, pp. 91–100, 1997. DOI: 10.1287/isre.12.1.83.9715 77

[17] U. Dayal, M. Castellanos, A. Simitsis, and K. Wilkinson. Data integration flows for business intelligence. In *Advances in Database Technology, Proc. 12th Int. Conf. on Extending Database Technology*, pp. 1–11, 2009. DOI: 10.1145/1516360.1516362 79

[18] C. E. Dyreson. Information Retrieval from an Incomplete Data Cube. In *Proc. 22th Int. Conf. on Very Large Data Bases*, pp. 532–543, 1996. 77

[19] C. E. Dyreson, T. B. Pedersen, and C. S. Jensen. Incomplete Information in Multidimensional Databases. In M. Rafanelli (ed.): *Multidimensional databases: Problems and Solutions*, pp. 282–309, Idea Group Publishing, 2003. 77

[20] R. Elmasri and S. Navathe. *Fundamentals of Database Systems*. 5th Edition. Addison Wesley, 2007. 75

[21] A. Eisenberg and J. Melton. SQL Standardization: The Next Steps. *ACM SIGMOD Rec.* 29(1):63–67, 2000. DOI: 10.1145/344788.344819 5, 27

[22] H. Garcia-Molina, J. D. Ullman, and J. Widom. *Database Systems*. 2nd Edition. Pearson Prentice Hall, 2009. 76

[23] M. Golfarelli. Open Source BI Platforms: A Functional and Architectural Comparison. In *Proc. 11th Int. Conf. Data Warehousing and Knowledge Discovery*, pp. 287–297. DOI: 10.1007/978-3-642-03730-6_23 80

[24] M. Golfarelli and S. Rizzi. *Data Warehouse Design: Modern Principles and Methodologies*. McGraw-Hill, 2009. 77

[25] M. Golfarelli, S. Rizzi. A Survey on Temporal Data Warehousing. *Int. J. Data Warehousing and Mining*, 5(1): 1–17, 2009. 78

[26] M. Golfarelli, S. Rizzi, and B. Vrdoljak. Data warehouse design from XML sources. In *Proc. ACM 4th Int. Workshop on Data Warehousing and OLAP*, pp. 40–47, 2001. DOI: 10.1145/512236.512242 80

[27] L. I. Gómez, B.Kuijpers, B. Moelans, A. A. Vaisman. A Survey of Spatio-Temporal Data Warehousing. In *Int. J. Data Warehousing and Mining*, 5(3):28–55, 2009. 78

[28] J. Gray, S. Chaudhuri, A. Bosworth, A. Layman, D. Reichart, M. Venkatrao, F. Pellow, and H. Pirahesh. Data Cube: A Relational Aggregation Operator Generalizing Group-By, Cross-Tab and Sub-Totals. *Data Mining and Knowledge Discovery*, 1(1):29–54, 1997. DOI: 10.1023/A:1009726021843 5, 25, 77

[29] M. Gyssens and L. V. S. Lakshmanan. A Foundation for Multi-Dimensional Databases. In *Proc. 23th Int. Conf. on Very Large Data Bases*, pp. 106–115, 1997. 77

[30] A. Y. Halevy, A. Rajaraman, and J. J. Ordille. Data Integration: The Teenage Years. In *Proc. 32nd Int. Conf. on Very Large Data Bases*, pp. 9–16, 2006. 79

[31] V. Harinarayan, A. Rajaraman, and J. D. Ullman. Implementing Data Cubes Efficiently. In *Proc. ACM SIGMOD Int. Conf. on Management of Data*, pp, 205–216, 1996. DOI: 10.1145/235968.233333 61

[32] S. Harinath, R. Zare, S. Meenakshisundaram, M. Carroll, and D.-Y. Lee. *Professional Microsoft SQL Server Analysis Services 2008 with MDX*. Wiley, 2009. 75

[33] J. A. Hoffer, M. B. Prescott, and and F. R. McFadden. *Modern Database Management*. 8th Edition. Pearson Prentice Hall, 2007. 75

[34] C. A. Hurtado, C. Gutiérrez and A. O. Mendelzon. Capturing summarizability with integrity constraints in OLAP. *ACM Trans. Database Syst.*, 30(3):854–886, 2005. DOI: 10.1145/1093382.1093388 77

[35] W. H. Immon. *Building the Data Warehouse*. Wiley, 1992. 14, 73

[36] W. H. Inmon. Data Mart Does Not Equal Data Warehouse. In *DM Review*, May, 1998. 17, 74

[37] W. H. Inmon. *Building the Operational Data Store*. 2nd Edition. Wiley, 1999. 74

[38] W. H. Inmon and A. Nesavich. *Tapping into Unstructured Data: Integrating Unstructured Data and Textual Analytics into Business Intelligence*. Prentice Hall, 2007. 74

[39] W. H. Inmon, D. Strauss, and G. Neushloss. *DW 2.0: The Architecture for the Next Generation of Data Warehousing*. Morgan Kaufman, 2008. 74

[40] International Standards Organisation (ISO). *ISO/IEC 9075-1:1999/Amd 1:2001 : AMDT 1: On-Line Analytical Processing (SQL/OLAP) - Information technology - Database languages - SQL - Part 1: Framework (SQL/Framework)*. ISO, 2001. 27

[41] H. V. Jagadish, L. V. S. Lakshmanan, and D. Srivastava. What can hierarchies do for data warehouses? In *Proc. 25th Int. Conf. on Very Large Data Bases*, pp. 530–541, 1999. 77

[42] C. S. Jensen, A. Kligys, T. B. Pedersen, and I. Timko. Multidimensional Data Modeling For Location-Based Services. *The VLDB J.*, 13(1):1–21, 2004. DOI: 10.1007/s00778-003-0091-3 79

[43] C. S. Jensen and R. T. Snodgrass. Temporal Data Management. In *IEEE Trans. Knowl. and Data Eng.*, 11(1):36–44, 1999. DOI: 10.1109/69.755613 78

[44] C. S. Jensen and R. T. Snodgrass. Semantics of Time-Varying Information. In *Inf. Syst.*, 21(4):311–352, 1996. DOI: 10.1016/0306-4379(96)00017-8 78

[45] M. R. Jensen, T. H. Møller, and T. B. Pedersen. Specifying OLAP Cubes on XML Data. *J. Intelligent Information Systems*, 17(2/3):255–280, 2001. DOI: 10.1023/A:1012814015209 80

[46] M. Kifer, A. Bernstein, and P. M. Lewis. *Database Systems—An Application-Oriented Approach*. 2nd Edition. Addison Wesley, 2006. 76

[47] R. Kimball and J. Casert. *The Data Warehouse ETL Toolkit: Practical Techniques for Extracting, Cleaning, Conforming, and Delivering Data*. Wiley, 2004. 74

[48] R. Kimball and M. Ross. *The Data Warehouse Toolkit*, 2nd Edition. Wiley, 2002. 2, 8, 13, 16, 17, 18, 35, 42, 74

[49] R. Kimball, M. Ross, W. Thornthwaite, and J. Mundy. *Kimball Group Reader: Relentlessly Practical Tools for Data Warehousing and Business Intelligence*. Wiley, 2010 74

[50] R. Kimball, M. Ross, W. Thornthwaite, J. Mundy, and B. Becker. *The Data Warehouse Lifecycle Toolkit*, 2nd Edition. Wiley, 2008. 17, 18, 74

[51] D. M. Kroenke. *Database Processing—Fundamentals, Design, and Implementation*. Pearson Prentice Hall, 2006. 76

[52] W. Lehner. Modeling Large Scale OLAP Scenarios. In *Advances in Database Technology, Proc. 6th Int. Conf. on Extending Database Technology*, pp. 153–167, 1998. DOI: 10.1007/BFb0100983 77

[53] H. Lenz and A. Shoshani. Summarizability in OLAP and Statistical Data Bases. In *Proc. 9th Int. Conf. on Scientific and Statistical Database Management*, pp. 39–48, 1997. DOI: 10.1109/SSDM.1997.621175 77

[54] C. Li and X. S. Wang. A Data Model for Supporting On-Line Analytical Processing. In *Proc. 5th Int. Conf. on Information and Knowledge Management*, pp. 81–88, 1996. DOI: 10.1145/238355.238444 77

[55] G. Luo, J. F. Naughton, C. J. Ellmann, and M. W. Waltzke. Transaction Reordering and Grouping for Continuous Data Loading. In *Proc. 1st Int. Workshop on Business Intelligence for the Real-Time Enterprises* pp. 34–49, 2006. DOI: 10.1007/978-3-540-73950-0_4 79

[56] E. Malinkowski and E. Zimányi. *Advanced Data Warehouse Design: From Conventional to Spatial and Temporal Applications.* Springer, 2009. 77, 79

[57] A. O. Mendelzon and A. A. Vaisman. Temporal Queries in OLAP. In *Proc. 26th Int. Conf. on Very Large Data Bases*, pp. 242–253, 2000. 78

[58] A. O. Mendelzon and A. A. Vaisman. Time in Multidimensional Databases. In M. Rafanelli (ed.): *Multidimensional Databases: Problems and Solutions*, pp. 166–199, Idea Group, 2003. 78

[59] Microsoft Corp. and Hyperion Solutions Corp. XML for Analysis Specification. `http://xmla.org`, 2001. 80

[60] V. Nebot, R. Berlanga, J. M. Pérez, M. J. Aramburu, and T. B. Pedersen. Multidimensional Integrated Ontologies: A Framework for Designing Semantic Data Warehouses. *LNCS Journal on Data Semantics - Special Issue on Semantic Data Warehouses* XIII, 2009. DOI: 10.1007/978-3-642-03098-7_1 80

[61] T. Niemi, M. Niinimaki, J. Nummenmaa, and P. Thanisch. Constructing an OLAP Cube from Distributed XML Data. *Proc. ACM 5th Int. Workshop on Data Warehousing and OLAP.*, pp. 22–37, 2002. DOI: 10.1145/583890.583894 80

[62] Oracle Corporation. Query Optimization in Oracle Database10g Release 2. `http://www.oracle.com/technology/products/bi/db/10g/pdf/twp_general_query_optimization_10gr2_0605.pdf`. Current as of July 19, 2010. 67

[63] D. Pedersen, K. Riis, and T. B. Pedersen. XML-Extended OLAP Querying. In *Proc. 14th Int. Conf. on Scientific and Statistical Database Management*, pp. 195–206, 2002. DOI: 10.1109/SSDM.2002.1029721 80

[64] T. B. Pedersen, J. Gu, A. Shoshani, and C. S. Jensen. Object-extended OLAP querying. *Data & Knowl. Eng.*, 68(5):453–480, 2009. DOI: 10.1016/j.datak.2008.10.008 79

[65] T. B. Pedersen and C. S. Jensen. Multidimensional Data Modeling for Complex Data. In *Proc. 15th Int. Conf. on Data Engineering*, pp. 336–345, 1999. 76, 77, 78

[66] T. B. Pedersen, C. S. Jensen, and C. E. Dyreson. Supporting Imprecision in Multidimensional Databases Using Granularities. In *Proc. 11th Int. Conf. on Scientific and Statistical Database Management*, pp. 90–101, 1999. DOI: 10.1109/SSDM.1999.787624 77

[67] T. B. Pedersen, C. S. Jensen, and C. E. Dyreson. Extending Practical Pre-Aggregation in On-Line Analytical Processing. In *Proc. 25th Int. Conf. on Very Large Data Bases*, pp. 663–674, 1999. 47

[68] T. B. Pedersen, C. S. Jensen, and C. E. Dyreson. A Foundation for Capturing and Querying Complex Multidimensional Data. *Inf. Syst.*, 26(5):383–423, 2001. DOI: 10.1016/S0306-4379(01)00023-0 11, 76, 77, 78

[69] T. B. Pedersen, D. Pedersen, and J. Pedersen. Integrating XML data in the TARGIT OLAP system. *Int. J. Web Engineering and Technology* 4(4):495–533, 2008. DOI: 10.1504/IJWET.2008.019945 80

[70] N. Pendse. What is OLAP? `http://www.bi-verdict.com/fileadmin/FreeAnalyses/fasmi.htm`. Current as of August 21, 2010. 2

[71] J. M. Pérez, R. Berlanga, M. J. Aramburu, and T. B. Pedersen, A relevance-extended multidimensional model for a data warehouse contextualized with documents. In *Proceedings of the Eighth ACM International Workshop on Data warehousing and OLAP*, pp. 19–28, 2005. DOI: 10.1145/1097002.1097007 80

[72] J. M. Pérez-Martínez, R. Berlanga-Llavori, M. J. Aramburu-Cabo, and T. B. Pedersen. Contextualizing data warehouses with documents. *Decision Support Systems* 45(1):77–94, 2008. DOI: 10.1016/j.dss.2006.12.005 80

[73] J. M. Pérez, R. Berlanga, M. J. Aramburu, and T. B. Pedersen: Integrating Data Warehouses with Web Data: A Survey. *IEEE Trans. Knowl. Data Eng.* 20(7): 940-955 (2008) DOI: 10.1109/TKDE.2007.190746 79

[74] J. Pokorny. Modelling Stars Using XML. In *Proc. ACM 4th Int. Workshop on Data Warehousing and OLAP*, pp. 24–31, 2001. DOI: 10.1145/512236.512240 80

[75] M. Rafanelli (Ed.). *Multidimensional Databases: Problems and Solutions.* Idea Group, 2003. 73

[76] M. Rafanelli and A. Shoshani. STORM: A Statistical Object Representation Model. In *Proc. 5th Int. Conf. on Scientific and Statistical Database Management*, pp. 14–29, 1990. 77

[77] R. Ramakrishnan and J. Gehrke. *Database Management Systems.* 3rd Edition. McGraw Hill, 2003. 75, 76

[78] S. Rivest, Y. Bedard, and P. Marchand. Toward better suppport for spatial decision making: Defining the characteristics of spatial on-line analytical processing. *Geomatica*, 55(4):539-555, 2001. 78

[79] D. A. Schneider. Practical Considerations for Real-Time Business Intelligence. In *Proc. 1st Int. Workshop on Business Intelligence for the Real-Time Enterprises*, pp. 1–3, 2006. DOI: 10.1007/978-3-540-73950-0_1 79

[80] Arie Shoshani. Statistical Databases: Characteristics, Problems, and some Solutions. In *Proc. 8th Int. Conf. on Very Large Data Bases*, pp. 208–222, 1982. 73

[81] A. Silberschatz, H. F. Korth, and S. Sudarshan. *Database System Concepts*. 5th Edition. McGraw-Hill, 2006. 76

[82] G. Spofford, S. Harinath, C. Webb, D. H. Huang, and F. Civardi. *MDX-Solutions: With Microsoft SQL Server Analysis Services 2005 and Hyperion Essbase*. Wiley, 2006. 25, 27, 75

[83] R. Stackowiak, J. Rayman, and R. Greenwald. *Oracle Data Warehousing and Business Intelligence Solutions: With Business Intelligence Solutions*. Wiley, 2007. 75

[84] M. Thiele, A. Bader, and W. Lehner. Multi-Objective Scheduling for Real-Time Data Warehouses. In *Proc. 13th Datenbanksysteme in Business, Technologie und Web*, pp. 307–326, 2009. DOI: 10.1007/s00450-009-0062-z 79

[85] M. Thiele, U. Fischer, and W. Lehner. Partition-based workload scheduling in living data warehouse environments. *Inf. Syst.*, 34(4-5):382–399, 2009. DOI: 10.1016/j.is.2008.06.001 79

[86] C. Thomsen and T. B. Pedersen. A Survey of Open Source Tools for Business Intelligence. In *Int. J. Data Warehousing and Mining*, 5(3):56–75, 2009. DOI: 10.1007/11546849_8 80

[87] C. Thomsen, T. B. Pedersen, and W. Lehner. RiTE: Providing On-Demand Data for Right-Time Data Warehousing. In *Proc. 24th Int. Conf. on Data Engineering*, pp. 456–465, 2008. DOI: 10.1109/ICDE.2008.4497454 79

[88] E. Thomsen. *OLAP Solutions: Building Multidimensional Information Systems*. Wiley, 1997. 3, 8

[89] E. Thomsen, G. Spofford, and D. Chase. *Microsoft OLAP Solutions*. Wiley, 1999. 27, 75

[90] I. Timko, C. E. Dyreson, and T. B. Pedersen. Probabilistic Data Modeling and Querying for Location-Based Data Warehouses. In *Proc. 17th Int. Conf. on Scientific and Statistical Database Management*, pp. 273–282, 2005. 78

[91] A. A. Vaisman and A. O. Mendelzon. A Temporal Query Language for OLAP: Implementation and a Case Study. In *Proc. 8th Int. Workshop on Database Programming Languages*, pp. 78–96, 2001. DOI: 10.1007/3-540-46093-4_5 78

[92] P. Vassiliadis. Modeling Multidimensional Databases, Cubes, and Cube Operations. In *Proc. 10th Int. Conf. on Scientific and Statistical Database Management*, pp. 53–62, 1998. DOI: 10.1109/SSDM.1998.688111 77

[93] P. Vassiliadis and T. K. Sellis. A Survey of Logical Models for OLAP Databases. *ACM SIGMOD Rec.*, 28(4):64–69, 1999. DOI: 10.1145/344816.344869 77

[94] P. Westerman. *Data Warehousing: Using the Wal-Mart Model.* Morgan Kaufmann, 2000. 76

[95] J. Widom. Research Problems in Data Warehousing. In *Proc. 4th Int. Conf. on Information and Knowledge Management*, pp. 25–30, 1995. DOI: 10.1145/221270.221319 73

[96] X. Yin and T. B. Pedersen. What Can Hierarchies Do for Data Streams? In *Proc. 1st Int. Workshop on Business Intelligence for the Real-Time Enterprises*, pp. 4–19, 2006. DOI: 10.1007/978-3-540-73950-0_2 79

[97] T. Zurek and M. Sinnwell. Data Warehousing Has More Colours Than Just Black and White. In *Proc. 25th Int. Conf. on Very Large Data Bases*, pp. 726–729, 1999. 77

Authors' Biographies

CHRISTIAN S. JENSEN

Christian S. Jensen, Ph.D., Dr.Techn., is a Professor of Computer Science at Aarhus University, Denmark, and a part-time Professor at Aalborg University, Denmark. Prior to joining Aarhus in 2010, he was employed full-time at Aalborg University. From September 2008 to August 2009, he was on sabbatical at Google Inc., Mountain View. He has previously been on several sabbaticals at University of Arizona.

His research concerns data management and spans semantics, modeling, indexing, and query and update processing. During the past decade, his main focus has been on topics within spatio-temporal data management.

He is an IEEE Fellow, a member of Royal Danish Academy of Sciences and Letters, the Danish Danish Academy of Technical Sciences, and the EDBT Endowment; and he is a trustee emeritus of the VLDB Endowment. In 2001 and 2002, he received Ib Henriksen's Research Award for his research in mainly temporal data management and Telenor's Nordic Research Award for his research in mobile services and data management.

He is Vice-Chair of ACM SIGMOD. He is an editor-in-chief of the VLDB Journal and has served on the editorial boards of ACM TODS, IEEE TKDE, and the IEEE Data Engineering Bulletin. He was PC chair or co-chair for STDM 1999, SSTD 2001, EDBT 2002, VLDB 2005, MobiDE 2006, MDM 2007, TIME 2008, DMSN 2008, and ISA 2010. He will be PC chair of ACM SIGSPATIAL GIS 2011.

TORBEN BACH PEDERSEN

Torben Bach Pedersen, Ph.D., is a Professor of Computer Science at Aalborg University, Denmark. Prior to joining Aalborg University in 2000, he worked in the software industry for more than six years.

His research concerns business intelligence technologies such as multidimensional databases, multidimensional data modeling, data warehousing, on-line analytical processing, data mining, and data integration with a focus on complex settings such as (semantic) web data, spatio-temporal data, and location-based services. He collaborates actively with many companies in the business intelligence industry.

He is a member of the SSTD Endowment, ACM, and IEEE. In 2004, he received the Young Reseachers Award from the European Consortium of Innovate Universities for his research in business intelligence. He serves on the editorial boards of the International Journal of Data Warehousing and Mining, Journal of Computer Science and Engineering, and LNCS Transactions on Large-Scale Data- and Knowledge-Centered Systems. He was PC Chair for DOLAP 2007 and DaWaK 2009 and 2010. He served as General Chair for SSTD 2009. He has served on more than 60 program committees, including VLDB, ICDE, and EDBT.

CHRISTIAN THOMSEN

Christian Thomsen, Ph.D., is a Postdoctoral Fellow in Center for Data-Intensive Systems at the Department of Computer Science, Aalborg University, Denmark. In 2008, he received his Ph.D. in Computer Science from Aalborg University.

His research interests concern business intelligence and include multidimensional databases, data warehousing, right-time data warehousing, extract-transform-load processes, and online analytical processing. He has served on several program committees including DaWaK. He has given courses on data warehousing and related technologies to master's students and industrial practitioners. Further, he has organized business intelligence related seminars that span academia and industry.

Index

Lightning Source UK Ltd.
Milton Keynes UK
UKOW042158160212

187407UK00001B/26/P